NAME

Place Value

Tell what the 7 in each number means.

1. 176,432,995 ______
2. 205,733,109 ______
3. 7,521,689,123 ______

Which digits are in the millions period?

4. 234,126,789,289 ______
5. 12,609,698,367 ______

Which digits are in the billions period?

6. 283,170,233 ______
7. 5,109,274,847 ______

Write each number in words.

8. 6,123,489 ______

9. 1,234,698,275 ______

Write in standard form. Remember to use commas.

10. 5 billion, 235 million, 125 ______
11. 258 million, 143 thousand, 308 ______

Write each number in expanded form.

12. 2,569,137 ______
13. 59,209,362 ______
14. 278,625 ______

15. Circle the digit in the ten-millions place. 1,489,275,303
16. Circle the digit in the billions place. 456,239,127,299

NAME **P2**

SHARPEN YOUR SKILLS

Comparing and Ordering Whole Numbers

Compare these numbers. Use >, <, or =.

1. 3,446 ◯ 3,664

2. 25,239 ◯ 23,593

3. 123,089 ◯ 132,809

4. 507,358 ◯ 507,358

5. 2,056,133 ◯ 2,506,133

6. 99,274 ◯ 99,247

List the numbers in order from least to greatest.

7.	317	371	307	______
8.	4,879	4,768	4,458	______
9.	56,332	56,233	56,033	______
10.	5,505	5,055	5,550	______
11.	123,450	132,540	123,405	______
12.	3,407,231	3,470,231	3,704,123	______

First compare the numbers. Then use blue to color all sections where the answer is >. Use yellow to color all sections where the answer is <. You will then see the flag of Sweden.

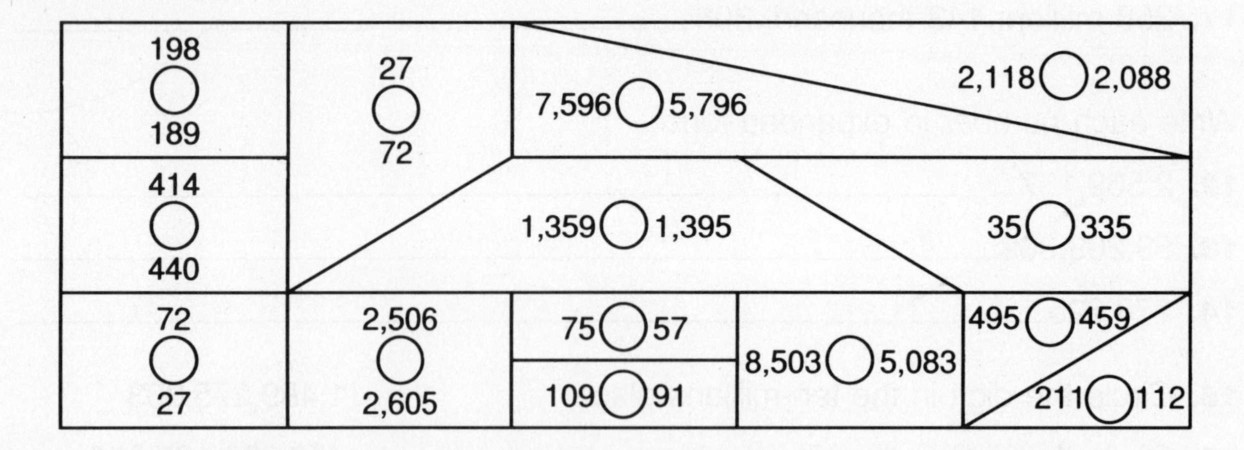

NAME

SHARPEN YOUR SKILLS

Rounding Whole Numbers

Round to the nearest ten.

1. 23 ________ **2.** 65 ________ **3.** 148 ________

4. 385 ________ **5.** 4,759 ________ **6.** 1,215 ________

Round to the nearest hundred.

7. 448 ________ **8.** 784 ________ **9.** 1,435 ________

10. 2,290 ________ **11.** 3,310 ________ **12.** 5,613 ________

Round to the nearest thousand.

13. 1,780 ________ **14.** 4,392 ________ **15.** 27,823 ________

Round to the nearest ten-thousand.

16. 14,657 ________ **17.** 28,861 ________ **18.** 8,964 ________

Round the numbers below to the nearest hundred. Then shade the sections where the answer is 800 or 1,300.

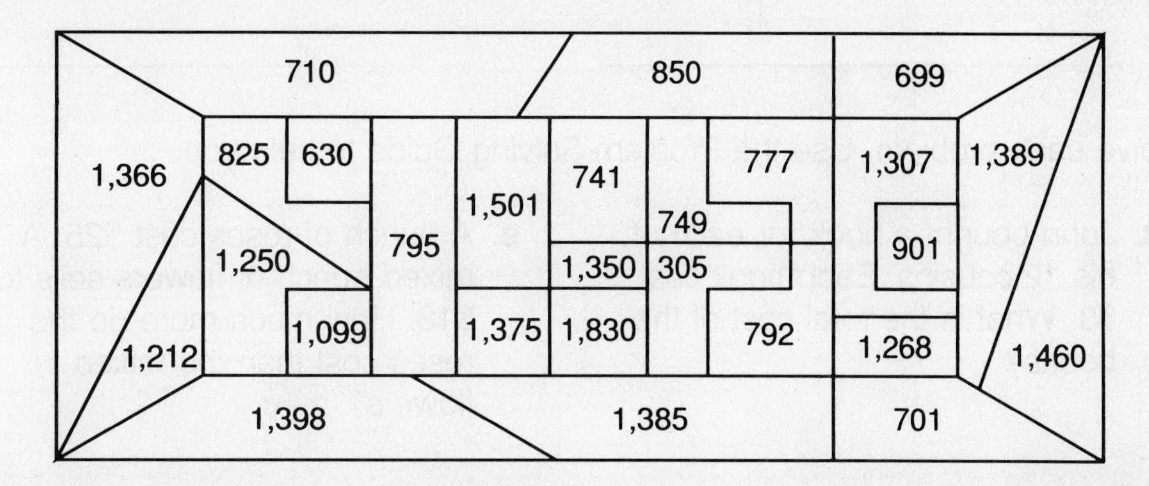

Using a Problem-Solving Guide

Answer the exercises about each problem below.

On a bicycle trip, Joan and her three friends rode 10 miles the first day, 9 miles the second day, and 5 miles the third day. How much farther did they bicycle on the first day than on the third day?

1. What are you asked to find?

____

____

2. What label will you use for your answer to show what the number means?

____

3. What can you do to solve this problem?

____

4. What is your answer?

____

An inflatable boat costs $68. A life jacket costs $35. A waterproof bag costs $42. How much will it cost to buy the life jacket and the boat?

5. What facts are given?

____

____

____

6. What facts are not needed to solve the problem?

____

7. Solve the problem. Give this answer in a sentence.

____

____

Solve each problem. Use the Problem-Solving Guide to help you.

8. John bought a book for each of his 12 cousins. Each book cost $3. What is the total cost of the books?

____

9. A bunch of roses cost $25. A mixed bunch of flowers sells for $18. How much more do the roses cost than the mixed flowers?

____

Use after pages 10–11.

SHARPEN YOUR SKILLS

Estimating Sums

Estimate each sum. Tell which strategy you used.

1. 2,234 + 1,389

2. 5,875 + 3,345

Estimate each sum by finding a range.

3. 385 + 639

4. 1,875 + 2,431

Estimate each sum by rounding both addends to the same place.

5. 298 + 312

6. 4,873 + 986

Estimate by comparing to a reference point.

7. 31 + 33 + 31

8. 59 + 58 + 59 + 57

Estimate the sum of each problem. Solve the problem by finding the actual answer. Use your estimate to explain if your answer is reasonable or not.

9. A newly built library received a shipment of 735 books. The next month, 189 books were delivered. How many books did the library receive in all?

10. A movie theater sold 148 tickets on Monday night and 695 tickets on Friday night. How many more tickets did the movie theater sell on Friday night?

NAME

SHARPEN YOUR SKILLS

Adding Whole Numbers

For each exercise estimate the sum. Then find the actual sum.

1. 58 + 24

2. 724 + 135

3. 528 + 42

4. 125 + 674

5. 78 + 18

6. 2,563 + 326

7. 684 + 225

8. 4,231 + 5,396

9. $23.40 + 47.38

10. $145.35 + 238.23

11. $509.36 + 63.61

12. 4,709 + 5,530

13. 1,008 + 876

14. 2,456 + 352 + 221

15. 1,480 + 487 + 604

16. 1,786 + 1,342 + 6,280

Solve each problem.

17. John studied many new words for a spelling test. One list he learned had 327 words. Another list had 952 words. How many new words did John learn?

____

18. On Monday 4,689 cars paid the toll as they crossed the bridge. On Tuesday, 3,209 cars paid the toll. How many cars in all paid the toll on both days?

____

Estimation The sum of which two numbers is closest to 5,000?

19. 1,300 3,560 2,800

____

20. 2,030 1,020 1,756

____

SHARPEN YOUR SKILLS

Mental Math for Addition

Find each sum using the mental math strategy of looking for numbers you can break apart.

1. 51 + 44 ________

2. 432 + 203 ________

3. 182 + 504 ________

4. 74 + 42 ________

5. 352 + 122 ________

6. 271 + 411 ________

Find each sum using the mental math strategy of combinations.

7. 7 + 6 + 13 + 4 ________

8. 74 + 83 + 17 ________

9. 93 + 41 + 7 ________

10. 8 + 77 + 2 ________

11. 97 + 173 + 3 ________

12. 4 + 160 + 96 ________

Use mental math to choose the correct sum.

13. 34 + 201

235 135 225

14. 12 + 22 + 88

112 122 132

15. 5 + 794 + 95

896 894 804

16. 65 + 99 + 35

199 299 109

17. 7 + 28 + 3 + 2

30 40 50

18. 21 + 86

106 107 117

NAME

Give Sensible Answers

Choose the most sensible answer.

1. Three classes went to the seashore for a field trip. How many students went?

10 100 1,000

2. How many hours did the students spend at the seashore that day?

5 50 500

3. The classes visited the information center to see a film about marine life. How many minutes did the film last?

2 20 200

4. Each student in Ms. Wing's class made a fish print to take home as a souvenir. How many fish prints were made by the class?

35 350 3,500

5. Mr. Fort's class walked along the shore toward the tide pools. How many miles did the class walk?

1 100 1,000

6. Mr. Garcia's class saw a fishing boat. There were tanks full of small shrimp. How many shrimp were there?

7 70 7,000

7. There were 64 people on a tour boat. How many of these people were crew members?

6 60 600

8. Jodi and Mark helped arrange a class picnic. How many cartons of milk were there if each student was allowed two cartons?

7 70 700

NAME

SHARPEN YOUR SKILLS

Adding Larger Numbers

For each exercise, estimate the sum. Then find the actual sum.

1. 3,456 + 6,274

2. 8,233 + 838

3. 72,910 + 13,894

4. 655 + 247 + 88

5. 4,474 + 3,843

6. 5,008 + 3,131

7. 145,772 + 27,165

8. 6,666 + 143

9. $407.85 + 332.08

10. $45.89 + 337.02

11. $119.92 + 90.05

12. $2,346.12 + 3,729.35

Estimate the sum using front-end digits with adjusting.

13. 154 + 316 + 259

Mental Math Use the mental math technique of compensation to find the sum.

14. 49 + 25

15. 152 + 599

NAME

SHARPEN YOUR SKILLS

Estimating Differences

Estimate each difference using front-end digits.

1. 4,256 − 2,345 ________

2. 8,678 − 5,532 ________

3. 6,359 − 5,781 ________

Estimate each difference by rounding both numbers to the same place.

4. 594 − 380 ________

5. 825 − 211 ________

6. 215 − 104 ________

7. 873 − 620 ________

8. 5,689 − 314 ________

9. 3,889 − 1,843 ________

10. 7,179 − 4,160 ________

11. 9,449 − 3,438 ________

12. 8,287 − 1,043 ________

13. 7,886 − 3,102 ________

14. 6,885 − 1,831 ________

15. 4,925 − 2,126 ________

Estimate the answer to this problem. Then solve the problem by finding the actual answer. Use your estimate to explain if your answer is reasonable.

16. A stationery store ordered 5,457 pages of lined paper and 3,424 pages of unlined paper. How many more pages of lined paper than unlined paper were ordered?

____

Use after pages 24–25.

NAME

SHARPEN YOUR SKILLS

Subtracting Whole Numbers

Estimate each difference. Then find the actual difference.

1. 24 − 19

2. 36 − 18

3. 46 − 27

4. 53 − 26

5. 795 − 57

6. 317 − 70

7. 285 − 92

8. 378 − 89

9. 522 − 128

10. 649 − 378

11. 219 − 175

12. 836 − 558

13. 8,762 − 6,958

14. 5,293 − 2,486

15. 9,536 − 3,681

16. 2,618 − 1,399

17. $23.67 − 18.38

18. $69.29 − 22.88

19. $125.33 − 66.21

20. $2,459 − 1,095

Solve the problem.

21. Wall School has 3,246 students, but 1,972 students will be transferred next term. How many students will remain at Wall School?

SHARPEN YOUR SKILLS

Mental Math for Subtraction

Find each difference using the mental math strategy of compensation.

1. 748 − 99

2. 356 − 98

3. 276 − 39

Find each difference using the mental math strategy of breaking apart numbers.

4. 85 − 34

5. 943 − 21

6. 374 − 52

Use mental math to find each difference. Tell which strategy you used.

7. 388 − 99

8. 297 − 271

9. $30.00 − $15.98

10. 185 − 71

11. 375 − 98

12. 853 − 42

Use after pages 28–29.

NAME

SHARPEN YOUR SKILLS

Subtracting Larger Numbers

For each exercise, estimate the difference. Then give the actual difference. **Remember** to check your work.

1. $\begin{array}{r} 78{,}642 \\ -\ 36{,}589 \\ \hline \end{array}$

2. $\begin{array}{r} 5{,}500 \\ -\ 2{,}396 \\ \hline \end{array}$

3. $\begin{array}{r} 6{,}283 \\ -\ 1{,}337 \\ \hline \end{array}$

4. $\begin{array}{r} 60{,}347 \\ -\ 54{,}628 \\ \hline \end{array}$

5. $\begin{array}{r} 44{,}000 \\ -\ 18{,}003 \\ \hline \end{array}$

6. $\begin{array}{r} 75{,}009 \\ -\ 32{,}115 \\ \hline \end{array}$

7. $\begin{array}{r} 9{,}800 \\ -\ 3{,}920 \\ \hline \end{array}$

8. $\begin{array}{r} 27{,}090 \\ -\ 11{,}107 \\ \hline \end{array}$

9. $\begin{array}{r} 58{,}102 \\ -\ 55{,}381 \\ \hline \end{array}$

10. 14,000 − 3,250 = ______

11. 20,098 − 6,126 = ______

12. 69,002 − 57,902 = ______

13. 37,000 − 16,013 = ______

Subtract across. Subtract down.

33,230	1,818	
752	269	

9,233	5,078	
4,673	3,899	

Use after pages 30–31.

SHARPEN YOUR SKILLS

Write an Equation

Write an equation. Then find the answer.

1. The article said 145 girls and 150 boys entered a spelling contest. How many children in all entered the contest?

____

2. Matthew's bike cost $225. Carol's bike cost $175. How much less did Carol pay than Matthew?

____

3. Last year 112 performers auditioned for the winter carnival. Only 75 were chosen. How many performers were not hired?

____

4. The blue team hiked 85 miles and the orange team hiked 37 miles less. How far did the orange team hike?

____

5. A total of 155 hamburgers were sold at the school barbecue. There also were 278 drinks sold. How many more drinks were sold than hamburgers?

____

6. A walking club clocked 364 miles the first year. The next year they clocked 452 miles. How many miles did they clock in all?

____

7. Esther drove 458 miles on the first day and 525 miles on the second day. How much farther did she drive on the second day?

____

8. In one week at school, 530 students arrived on time and 89 students were late. How many students attended school in all?

____

Use after pages 32–33.

NAME

SHARPEN YOUR SKILLS

Missing Addends

Find the missing addend. Use families of facts to help.

1. $28 + n = 68$

2. $25 + n = 63$

3. $74 + n = 82$

4. $121 = n + 18$

5. $281 = n + 198$

6. $801 = n + 372$

7. $654 + n = 6{,}578$

8. $1{,}713 = 820 + n$

9. $1{,}942 = n + 512$

10. $506 + n = 2{,}679$

11. $743 + n = 5{,}287$

12. $1{,}590 = n + 480$

Solve the problem.

13. An airline added 148 employees to its staff. The airline then had 1,375 employees. How many employees did the airline have to begin with?

SHARPEN YOUR SKILLS

Multiples

List 5 multiples of the following numbers.

1. 4 **2.** 5 **3.** 6 **4.** 7

5. 8 **6.** 9 **7.** 3 **8.** 2

List three common multiples of the following.

9. 3 and 6 **10.** 2 and 8 **11.** 4 and 6

12. 4 and 5 **13.** 5 and 6 **14.** 6 and 8

Find the least common multiple for these numbers.

15. 5 and 6 **16.** 6 and 7 **17.** 3 and 8

18. 3 and 7 **19.** 5 and 10 **20.** 4 and 9

Use after pages 46–47.

NAME

SHARPEN YOUR SKILLS

Mental Math for Multiples of 10, 100, and 1,000

Multiply. **Remember** to use mental math.

1. $6 \times 10 =$ ______ **2.** $14 \times 10 =$ ______ **3.** $10 \times 52 =$ ______

4. $200 \times 9 =$ ______ **5.** $50 \times 50 =$ ______ **6.** $4 \times 30 =$ ______

7. $900 \times 900 =$ ______ **8.** $1{,}000 \times 19 =$ ______

9. $20 \times 2{,}000 =$ ______ **10.** $400 \times 600 =$ ______

11. $250 \times 10 =$ ______ **12.** $5{,}000 \times 100 =$ ______

I feed dogs and cats and water plants. What is my career?

To find out, use a ruler to connect each exercise to its answer.
Each line will go through a letter. Write the letter in the blank.

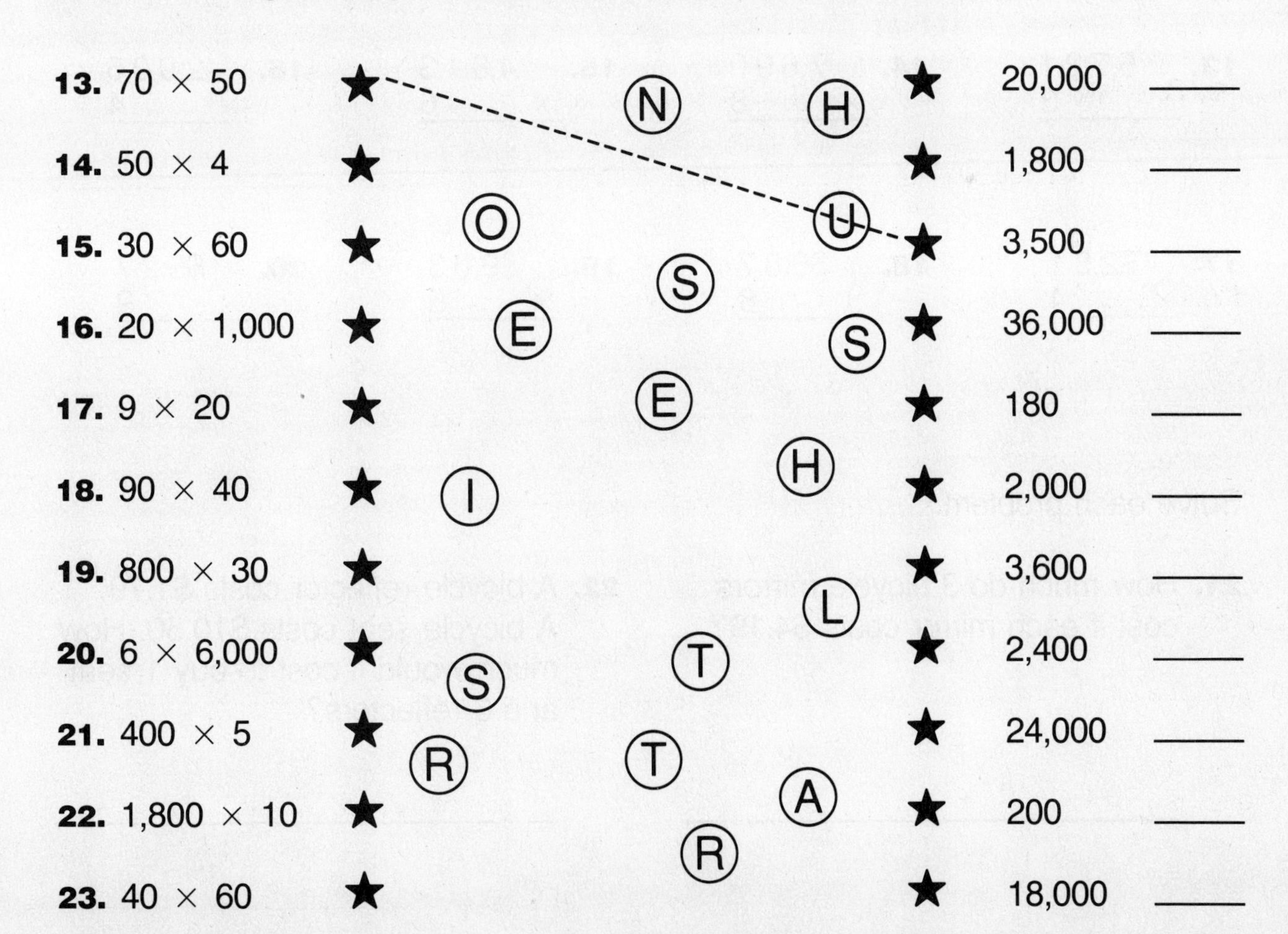

NAME

SHARPEN YOUR SKILLS

Multiplying by a One-Digit Number

Multiply. **Remember** to estimate to tell if the answer is reasonable.

1. 64×2

2. 72×3

3. 21×6

4. 50×8

5. 49×3

6. 58×9

7. 16×5

8. 34×7

9. 321×4

10. 503×7

11. 627×2

12. 809×6

13. $5{,}721 \times 3$

14. $7{,}690 \times 8$

15. $4{,}813 \times 6$

16. $2{,}905 \times 4$

17. $\$2.51 \times 4$

18. $\$6.37 \times 8$

19. $\$9.03 \times 5$

20. $\$5.27 \times 9$

Solve each problem.

21. How much do 3 bicycle mirrors cost if each mirror costs $4.19?

____

22. A bicycle reflector costs $1.79. A bicycle seat costs $10.50. How much would it cost to buy 1 seat and 3 reflectors?

____

NAME

SHARPEN YOUR SKILLS

Draw a Diagram

Jon can choose brown pants, blue pants, or black pants.
He can choose to wear a white or a tan shirt.

1. Complete the following tree diagrams to show all the possible choices of pants and shirts Jon has.
Then tell how many choices Jon has in all. ____________

Pants	Shirt	Choices
brown	white	brown pants, white shirt
	tan	brown pants, tan shirt
blue	white	____________
	tan	____________
black	____________	____________
	____________	____________

On another sheet of paper, make a tree diagram to help you list all possible choices. Then tell how many choices are available.

2. A group of students can play softball or kickball at school or at the park. How many choices are there?

____

3. A group of students could play soccer or football starting at 3:00, 4:00, or 5:00. How many choices are available?

____

4. The Don family could travel by plane, car, train, or bus to Chicago, Detroit, or New York. How many choices are available?

____

5. The Lee family wishes to visit either Idaho, Oregon, Utah, or Nevada in April or July. How many choices are available?

____

NAME

SHARPEN YOUR SKILLS

Multiplying by a Multiple of 10 or 100

Multiply. **Remember** to estimate to tell if the answer is reasonable.

1. 36×30

2. 63×40

3. 55×60

4. 95×60

5. 21×30

6. 78×20

7. 76×50

8. 82×30

9. 38×40

10. 77×60

11. 63×50

12. 15×70

13. 472×400

14. 896×600

15. 124×900

16. 697×200

17. $300 \times 895 =$ ______

18. $500 \times 907 =$ ______

19. $700 \times 215 =$ ______

20. $200 \times 307 =$ ______

Solve each problem.

21. A machine can stamp the date on 120 letters per minute. How many letters can it stamp in one hour?

______

22. An automatic sorter can sort 270 letters per minute. How many letters can it sort in one hour?

______

Use after pages 54–55.

NAME

SHARPEN YOUR SKILLS

Mental Math for Multiplication

Use mental math to find each product. Tell which strategy you used.

1. 31 × 6

2. 25 × 7 × 4

3. 7 × 67

4. 5 × 43 × 4

5. 5 × 299

6. 407 × 3

7. 3 × 52 × 10

8. 43 × 21

9. 103 × 12

10. 16 × 10 × 2

11. 44 × 3

12. 50 × 41 × 2

Solve each problem. **Remember** to use mental math when it is helpful.

13. Apples cost $3.05 a pound. How much will 6 pounds cost?

14. Avocados cost $1.09 each. How much will 12 avocados cost?

15. Carrots cost $0.33 per bag. How much will 12 bags cost?

16. Plums are $0.59 per pound. How much will 4 pounds cost?

NAME

SHARPEN YOUR SKILLS

Choosing a Computation Method

Tell which computation method you would use. Use *P* for pencil and paper, *M* for mental math, and *C* for a calculator. Then find each answer.

1. $6 \times 25 \times 4$ ________

2. 63×542 ________

3. $467 - 231$ ________

4. $507 + 312$ ________

5. $6 \times \$9.80$ ________

6. $8{,}102 - 743$ ________

7. $6 \times 7 \times 50$ ________

8. 40×70 ________

9. $\$14.39 \times 17$ ________

10. $150 \times 3 \times 2$ ________

11. $3{,}473 + 2{,}974$ ________

12. $199 - 50$ ________

Write a *P, M,* or *C* to tell which computation method you would use. Then find each answer.

NEWTOWN CLIPPERS

Box Seats $10.00

Reserved Lower Deck $8.00

Reserved Upper Deck $7.00

General Admission $3.00

13. A group wants 15 box seats. How much will the tickets cost?

____

14. Can a group buy 3 box seats, 2 reserved lower deck, 4 reserved upper deck, and 6 general admission seats for under $100?

____

15. The attendance for a weekend series to see the Clippers was

Friday	-	23,407
Saturday	-	34,291
Sunday	-	38,972

What was the total attendance?

____

16. What is the total cost of the tickets in Problem 14?

____

NAME

SHARPEN YOUR SKILLS

Estimating Products

Estimate each product. Write which strategy you used.

1. 37×5

2. 81×8

3. 29×2

4. 26×3

5. 76×4

6. 264×7

7. 388×3

8. 624×9

9. 78×66

10. 55×77

11. 45×38

12. 89×72

13. $4 \times 51 \times 3$

14. $24 \times 8 \times 2$

15. 504×21

16. 7×299

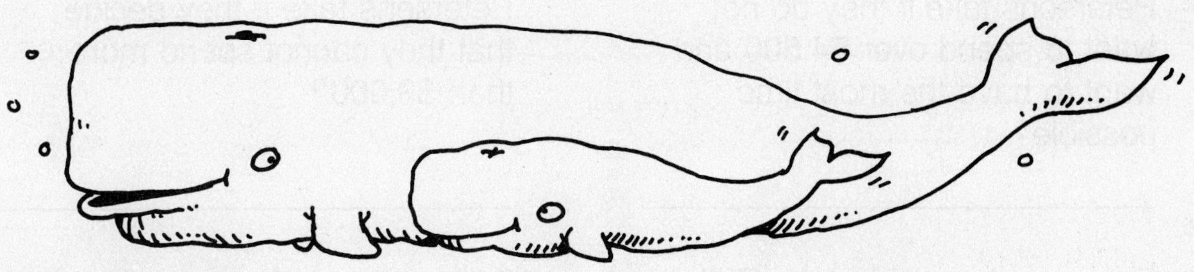

Solve the problem.

17. A whale and her calf swim 7 miles per hour. If they swim at this rate for 24 hours, estimate how far they will travel. Write the method you used.

SHARPEN YOUR SKILLS

Deciding When an Estimate Is Enough

Solve each problem.

A 7-day cruise costs $2,395 on Delta Lines; a 5-day cruise is $1,850. The Starry Night Lines offers a 7-day cruise for $2,135; a 5-day cruise is $1,695. On both lines children under 12 pay half price. Each price is per person.

1. A family of 4 wants to take the 5-day trip on the Starry Night. The children are 14 and 11. Can the whole family go for under $5,000.00? Use estimation.

2. What is the range of cost for the 3 adult fares for a 5-day cruise on the Starry Night Lines?

3. Kathy and Tony Peterson are going on their honeymoon. Give the estimated range of cost for the 5-day and 7-day trips on Delta Lines.

4. What is the estimated range of cost for the Petersens for the two trips on the Starry Night Lines?

5. Which cruise should the Petersons take if they do not want to spend over $4,500 and want to have the most time possible?

6. Which cruise should the Petersens take if they decide that they cannot spend more than $3,000?

7. Four people want to take Delta Line's 7-day cruise.They have $9,500. Will an estimate tell if they can afford it? Explain.

8. If these 4 people decide to take another cruise, which of the other three could they afford?

NAME

SHARPEN YOUR SKILLS

Multiplying by a Two-Digit Number

Find each product. Then shade in any boxes containing answers and find a path from START to STOP, using only shaded-in boxes. You may not need all of the boxes you shade in.

1. 62×33

2. 77×54

3. 20×46

4. 69×82

5. 300×29

6. 780×38

7. 997×23

8. 584×42

9. $36 \times 240 =$ ________

10. $28 \times 400 =$ ________

11. $31 \times 299 =$ ________

12. $15 \times 132 =$ ________

START 9,269	2,924	665	1,629	1,736
2,046	5,316	1,897	5,658	870
24,528	29,640	920	22,931	8,640
1,992	11,200	2,556	1,980	1,872
3,870	3,240	840	4,158	8,700 STOP

Use after pages 66–67.

SHARPEN YOUR SKILLS

Choose an Operation

Tell whether to add, subtract, or multiply.
Then solve the problem.

1. The Carters own a poultry farm. They had 982 chicks. They sold 426 of them. How many chicks did they have left?

____

____

2. Mr. Carter gathered 582 eggs on Monday, 606 eggs on Tuesday, and 594 eggs on Wednesday. How many eggs did he gather altogether?

____

____

3. The Scotts have a dairy farm with 125 cows in each pasture. If there are 5 pastures on the farm, how many cows do the Scotts own?

____

____

4. Mrs. Oliver uses about 275 liters of water on her farm each day. About how much water does she use in 28 days?

____

____

5. Mrs. Rice sold 45 hens in March. In April she sold 27 hens, in May she sold 118 hens, and in June she sold 152 hens. How many hens did she sell in the four months?

____

____

6. The Cavanaughs have 218 chickens, 35 cows, 18 sheep, and 9 pigs on their farm. How many animals do the Cavanaughs have altogether?

____

____

Use after pages 68–69.

NAME

SHARPEN YOUR SKILLS

Multiplying by a Three-Digit Number

Use paper and pencil or a calculator to multiply. **Remember** to estimate to tell if the answer is reasonable.

1. 250×354

2. 614×522

3. 400×354

4. 324×612

5. 872×300

6. 359×600

7. 846×500

8. 900×400

9. 750×220

10. 656×212

11. 797×151

12. 444×444

13. 672×274

14. 800×430

15. 924×296

16. 379×501

Solve the problem.

17. Each day, 850 people visit the penguin house at the zoo. How many people visit this exhibit in a year (365 days)?

____

NAME

SHARPEN YOUR SKILLS

Customary Units of Length

Choose the unit you would use for each measure.

1. Width of a book

inches feet

2. Height of a man

feet yards

3. Length of a garden

yards miles

4. Height of a school

inches feet

5. Distance between cities

yards miles

6. Length of a watchband

inches feet

Use a ruler. Measure to the nearest $\frac{1}{4}$ inch and to the nearest $\frac{1}{8}$ inch.

7. ____________

____

8. ____________

____

9. ____________

____

10. ____________

____

Use a ruler to draw a line of each length.

11. $1\frac{2}{8}$ inches

12. $3\frac{6}{8}$ inches

13. Longer than $1\frac{1}{2}$ inches and shorter than $2\frac{3}{8}$ inches

Use after pages 82–83.

NAME

SHARPEN YOUR SKILLS

Equal Customary Measures of Length

1 ft = 12 in.
1 yd = 3 ft
1 yd = 36 in.
1 mi = 5,280 ft
1 mi = 1,760 yd

Change each measure to the unit given.

1. 4 ft = __________ in.

2. 4 ft 7 in. = __________ in.

3. 2 ft = __________ in.

4. 1 ft 9 in. = __________ in.

5. 5 yd = __________ in.

6. 3 yd = __________ ft

7. 2 ft 11 in. = __________ in.

8. 3 mi = __________ ft

Choose the largest measure.

9. 6 ft 3 yd 50 in.

10. 2,000 yd 1 mi 4,000 ft

Solve each problem.

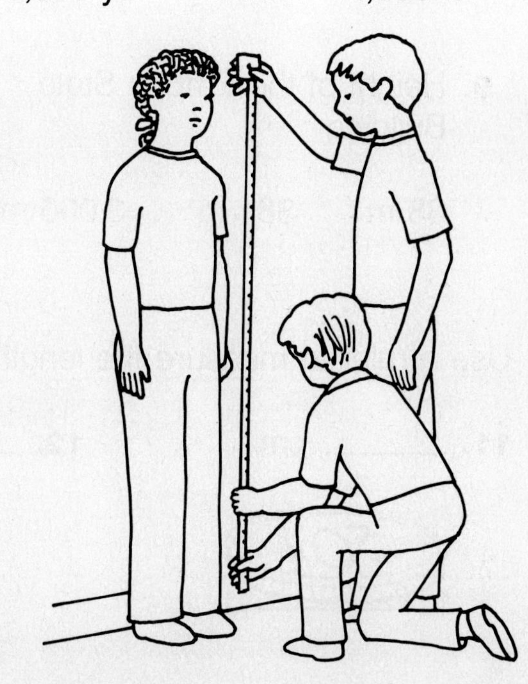

11. Brian is 3 ft 9 in. tall. How many inches is this?

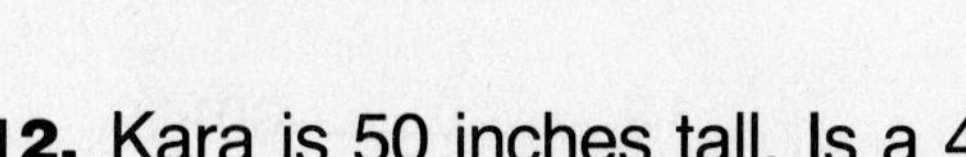

____

12. Kara is 50 inches tall. Is a 4-foot tape measure long enough to measure Kara's height?

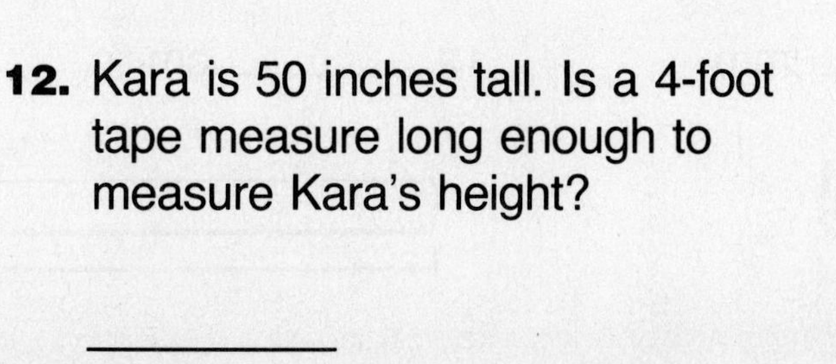

____

NAME

SHARPEN YOUR SKILLS

Metric Units of Length

Estimation Tell whether you would use *millimeters, centimeters, meters,* or *kilometers* to measure these.

1. Distance from Dallas, Texas, to Indianapolis, Indiana

____

2. Length of a basketball court

____

3. Width of a quarter

____

4. Width of Lake Superior

____

5. Length of a swimming pool

____

6. Length of this book

____

Choose the most sensible measure.

7. Width of a snowflake

2 mm 2 km 2 dm

8. Height of a tree

15 cm 15 m 15 km

9. Height of the Empire State Building

38 m 381 m 3,000 m

10. Length of a strand of spaghetti

22 mm 22 cm 22 m

Use a ruler to measure the length of each item.

11. _______ cm

12. _______ mm

13. _______ cm

SHARPEN YOUR SKILLS

Equal Metric Measures of Length

Write each measure in millimeters.

1 cm = 10 mm
1 dm = 10 cm
1 dm = 100 mm
1 m = 100 cm
1 m = 1,000 mm
1 km = 1,000 m

1. 5 cm = ______________ mm

2. 13 cm = ______________ mm

3. 12 m = ______________ mm

4. 5 m = ______________ mm

5. 4 km = ______________ mm

6. 3 dm = ______________ mm

Write each measure in centimeters.

7. 11 m = ______________ cm

8. 117 cm = ______________ mm

9. 2 m = ______________ cm

10. 25 cm = ______________ mm

Number Sense Tell which of these is the largest measure.

11. 2 dm 2 cm 2 m

12. 5 m 50 mm 10 dm

13. 12 cm 3 dm 120 mm

14. 45 m 10 dm 500 cm

Solve each problem.

15. The classroom ceiling is 3 meters high. How many centimeters is this?

____

16. A length of cable is 2 km long. How many meters is this?

____

Perimeter

Find the perimeter of each figure. **Remember** to label the unit of measure.

1.

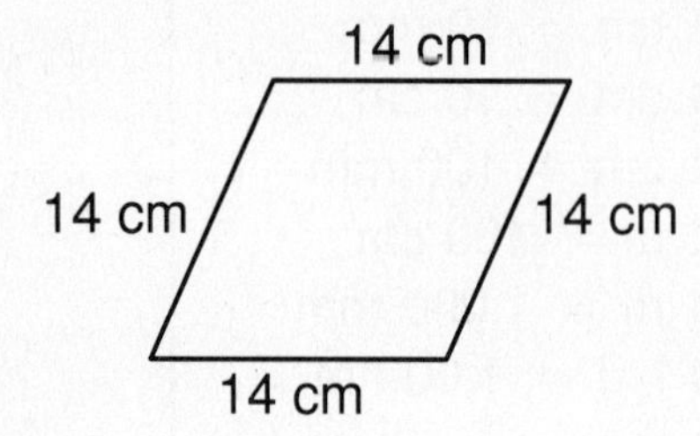

____

2.

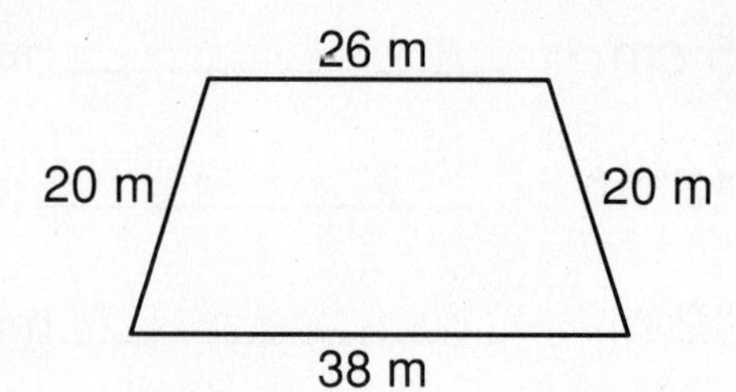

____

3.

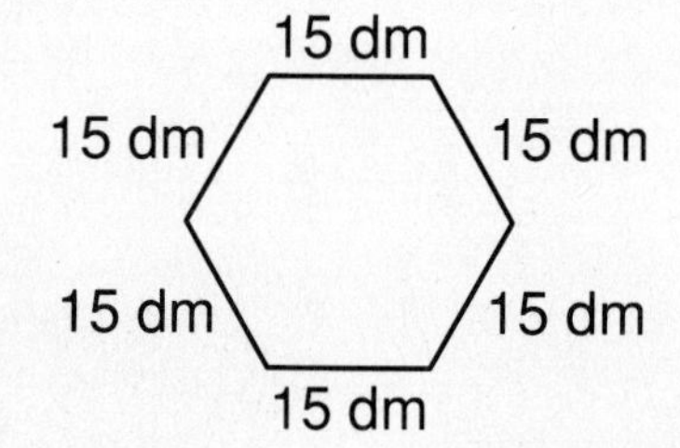

____

4.

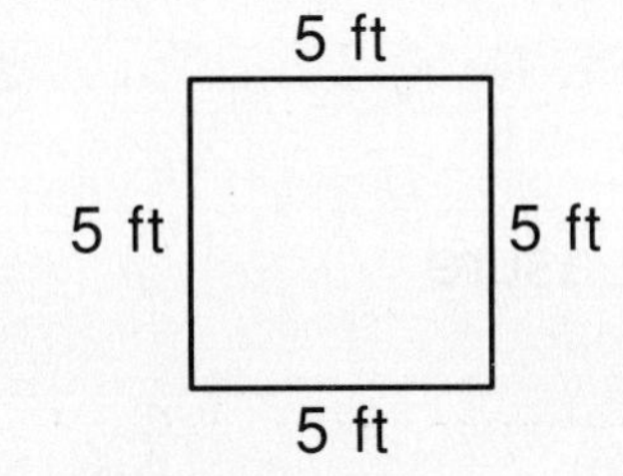

____

5.

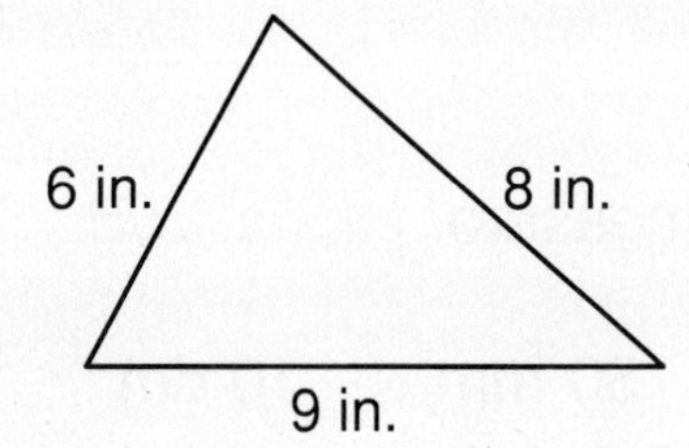

____

6.

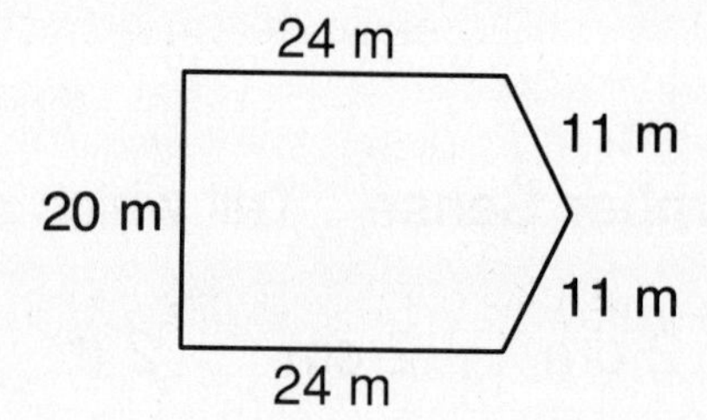

____

Solve the problem.

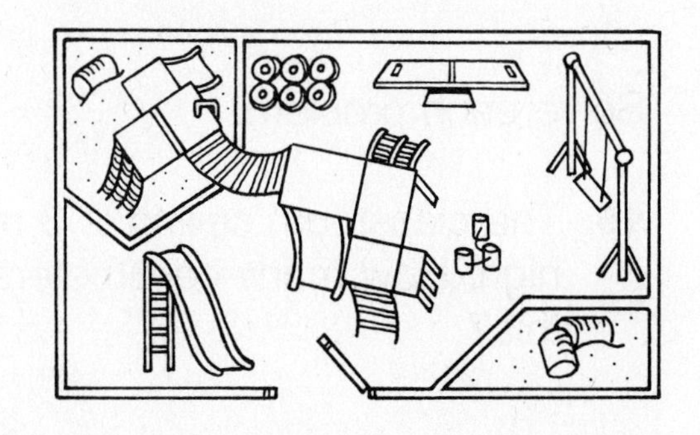

7. A rectangular playground has 4 sides that measure 33 meters, 25 meters, 33 meters, and 25 meters. How long is the fence that borders the playground?

____

Finding Area by Counting Squares

Find the area of each figure.
Each square is 1 cm².

1. 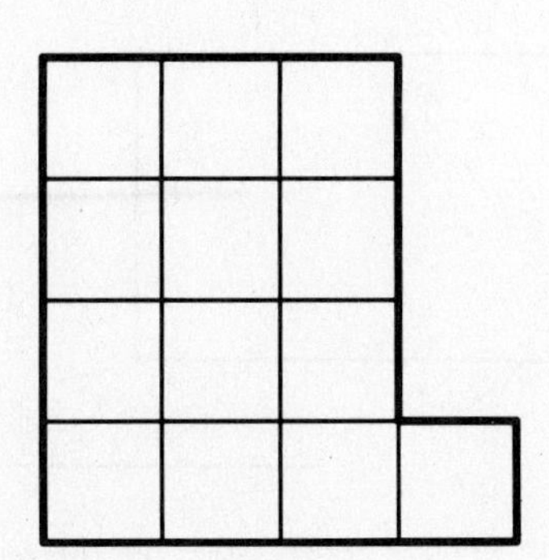

____

2. 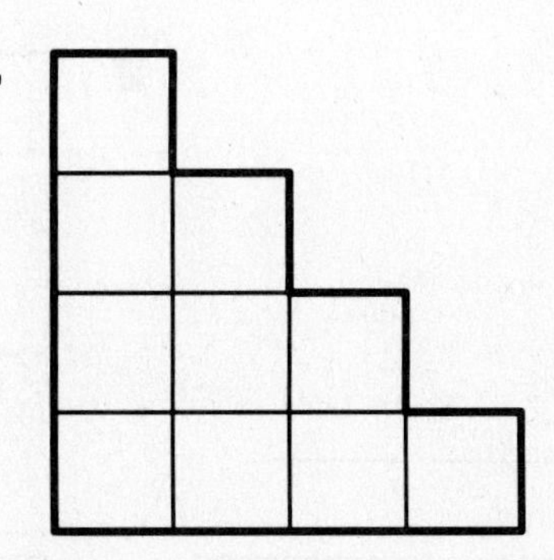

____

3. 

____

4. 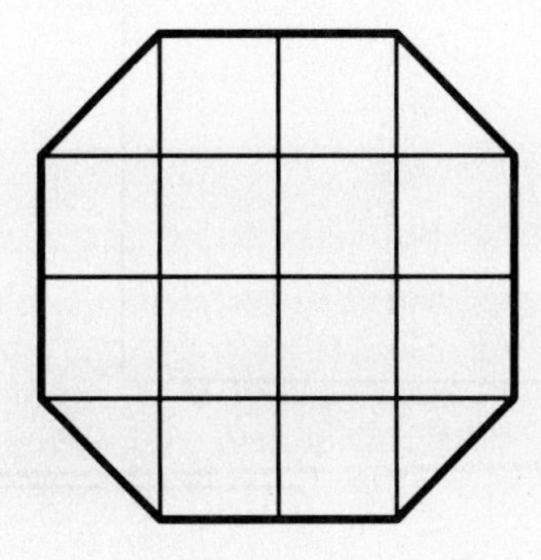

____

5. 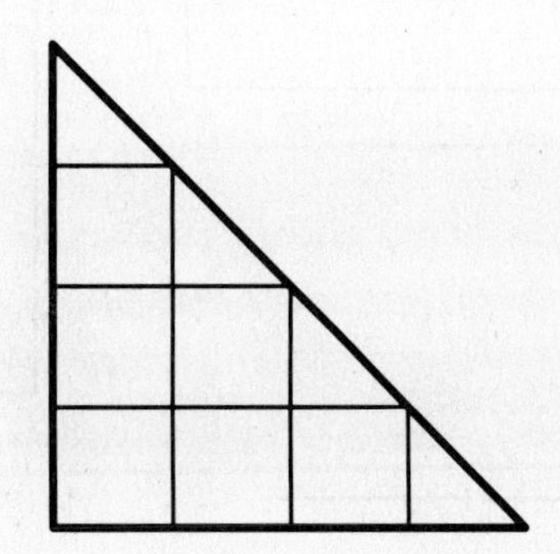

____

6. 

____

Find the area of each figure. Each square is 1 sq in.

7. 

____

8. 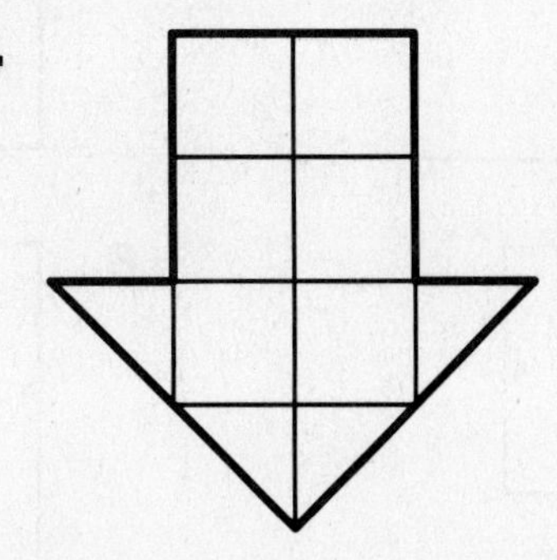

____

9. 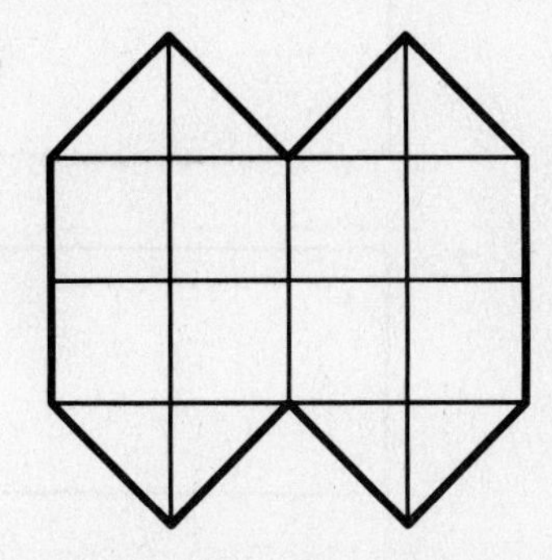

____

NAME

SHARPEN YOUR SKILLS

Area of Squares and Rectangles

Measure the length and width of each rectangle in inches. Then find the area of the figure.

1.

______

2.

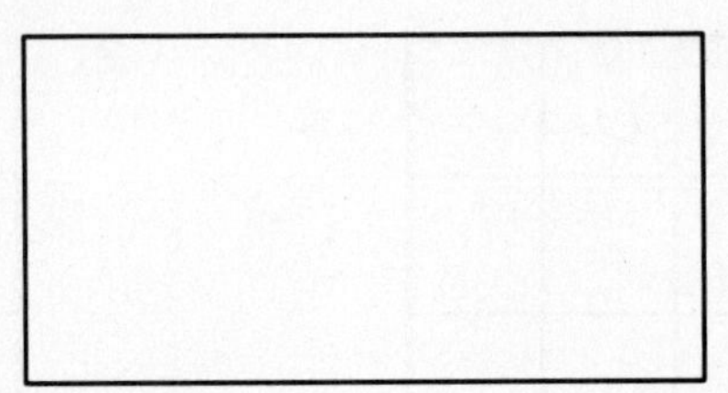

______

3.

______

4.

______

Measure the length and width of each rectangle in centimeters. Then find the area.

5.

______

6.

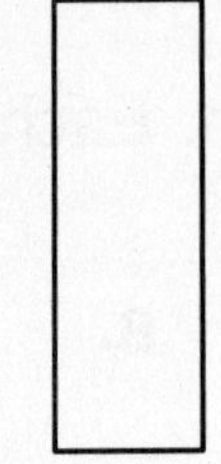

______

7.

______

8.

______

NAME

SHARPEN YOUR SKILLS

Make a Table

1. Jamie has 36 floor tiles, each 1 foot square. Complete the table to show the different rectangles he can make with the tiles.

Width (ft)	Length (ft)	Perimeter (ft)	Area (ft)
1	36	1 + 36 + 1 + 36 = _____	36
			36
			36
			36
			36

Solve the problems using the table.

2. What width and length of the tiles make the largest perimeter? What is the perimeter?

____

3. What width and length of the tiles make the smallest perimeter? What is the perimeter?

____

4. Jamie's closet is a square with a perimeter of 24 feet. How can he arrange the tiles to cover the floor?

____

5. Jamie's hallway is 3 feet wide and 10 feet long. Can he cover the entire hallway? How many tiles will be left over?

____

Use after pages 98–99.

SHARPEN YOUR SKILLS

Area of Triangles

Find the area of each triangle.
Remember to label each unit of measure.

1.

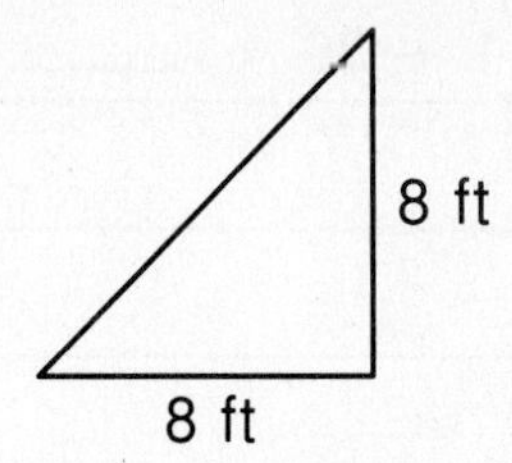

2.

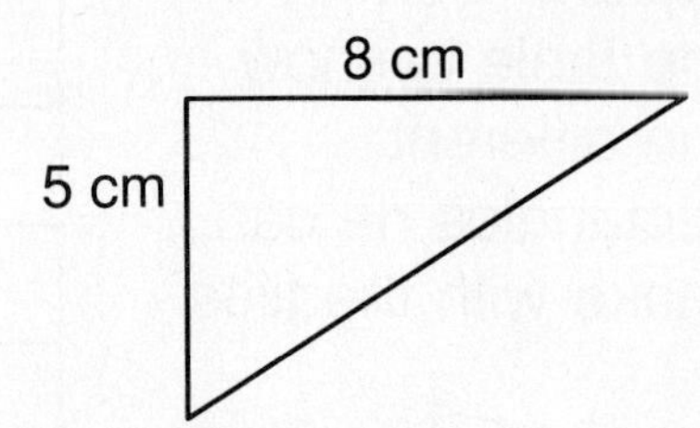

3.

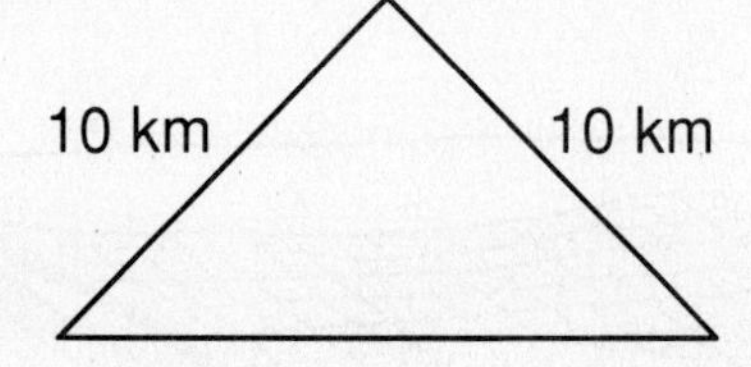

4.

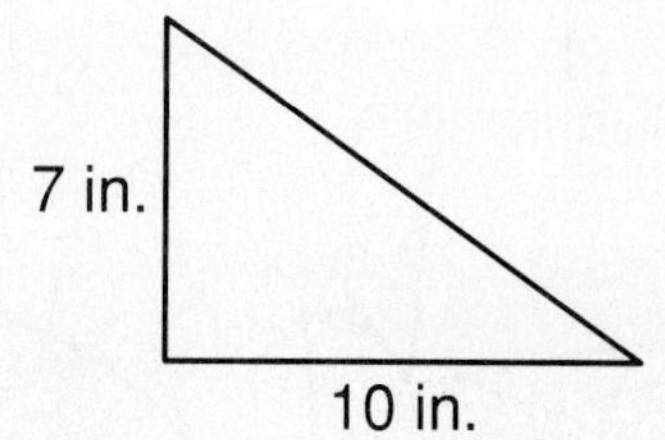

5.

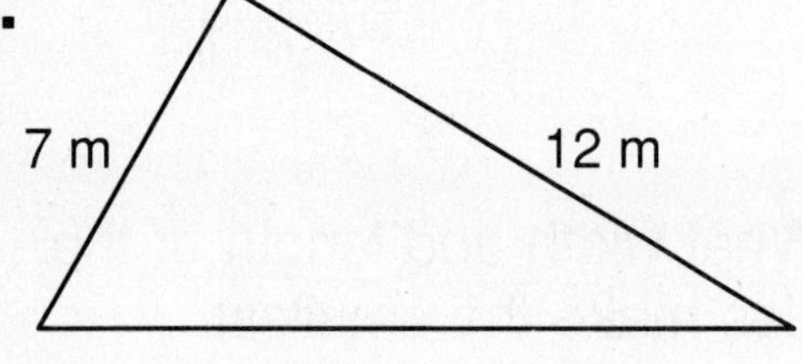

6.

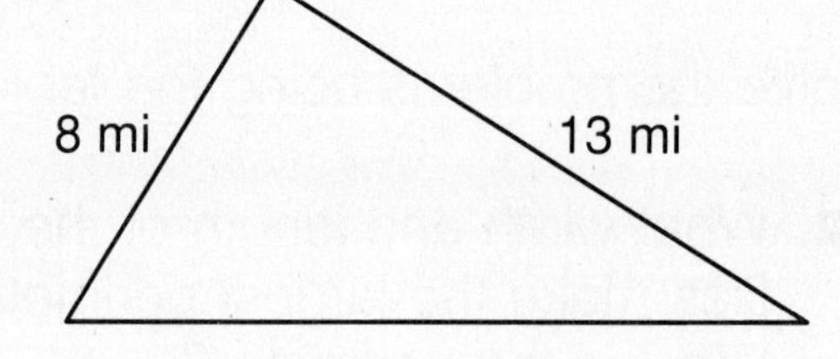

7.

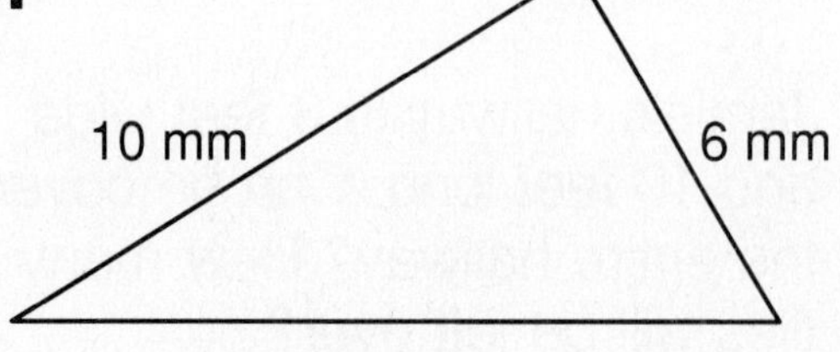

8.

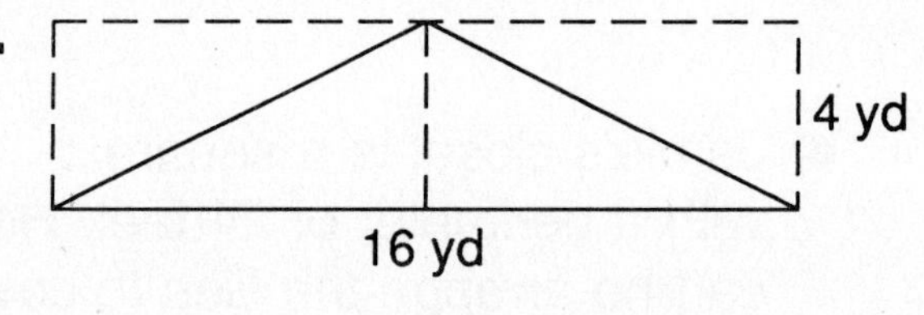

SHARPEN YOUR SKILLS

Use Data from a Picture

Find the area of each figure. **Remember** to label each unit of measure.

1.

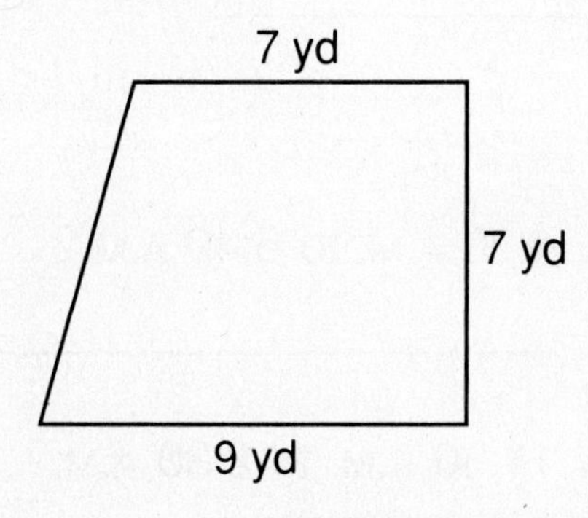

2.

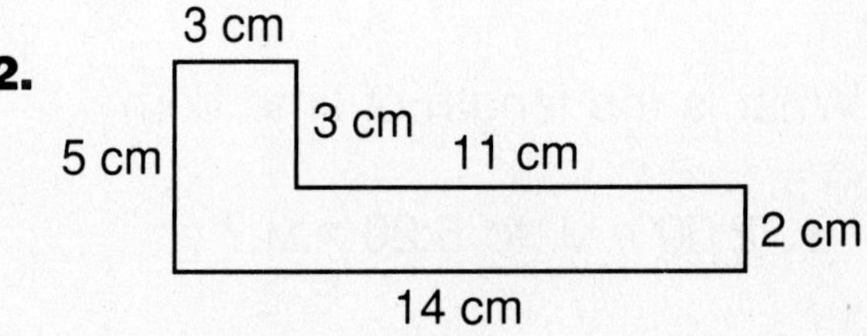

3.

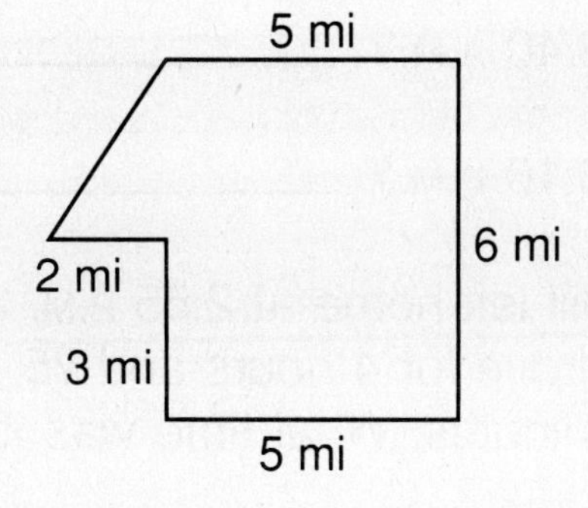

4.

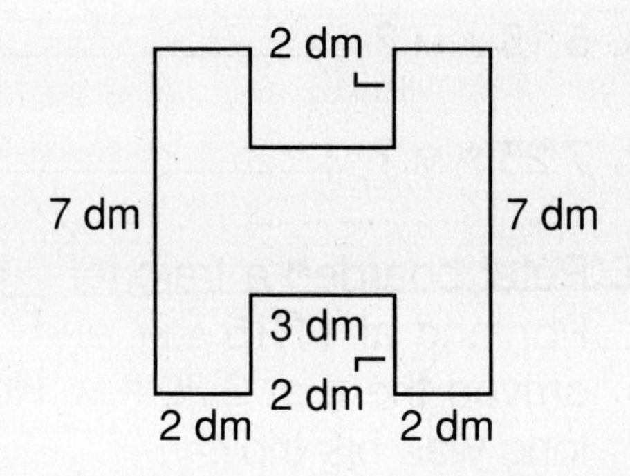

5.

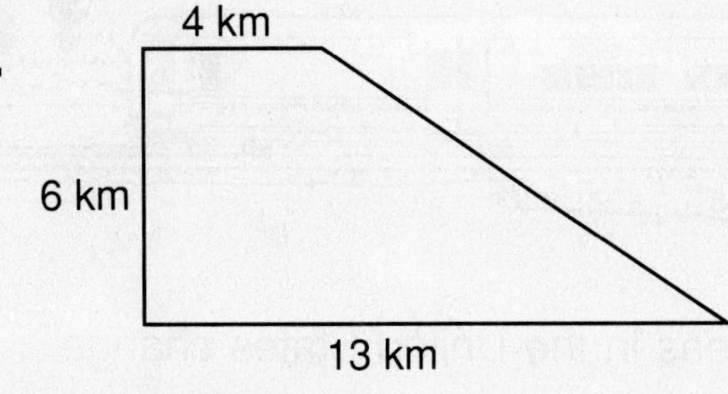

6.

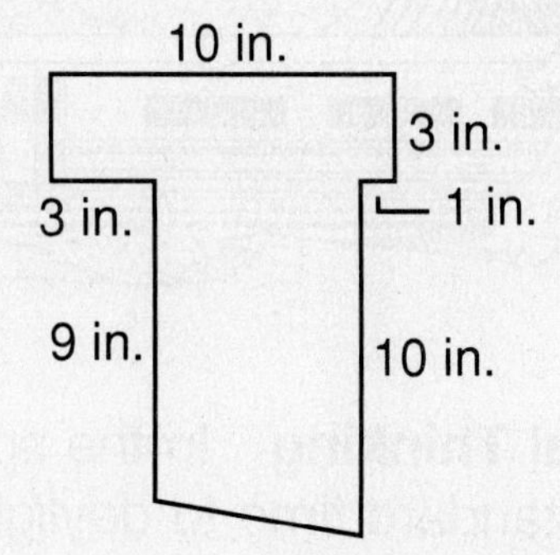

NAME

SHARPEN YOUR SKILLS

Time

1 min = 60 s	1 hr = 60 min	1 day = 24 h

What is the length of time from

1. 2:00 P.M. to 5:20 P.M.? ______
2. 4:20 A.M. to 5:40 A.M.? ______
3. 10:50 A.M. to 8:00 P.M.? ______
4. 11:30 P.M. to 2:30 A.M.? ______

What time will it be 3 hours and 20 minutes after

5. 5:15 A.M.? ______
6. 8:40 A.M.? ______
7. 7:25 P.M.? ______
8. 9:10 P.M.? ______
9. Peter boarded a train for Portland at 11:15 A.M. and arrived there at 2:20 P.M. How long was his trip? ______
10. Jill left home at 2:55 P.M. and drove for 4 hours and 25 minutes. What time was it then? ______

Critical Thinking In the spring, most areas in the United States change from standard time to daylight savings time. Why do you think standard time resumes in the fall?

______

NAME

SHARPEN YOUR SKILLS

Temperature: Celsius and Fahrenheit

Choose the more sensible temperature.

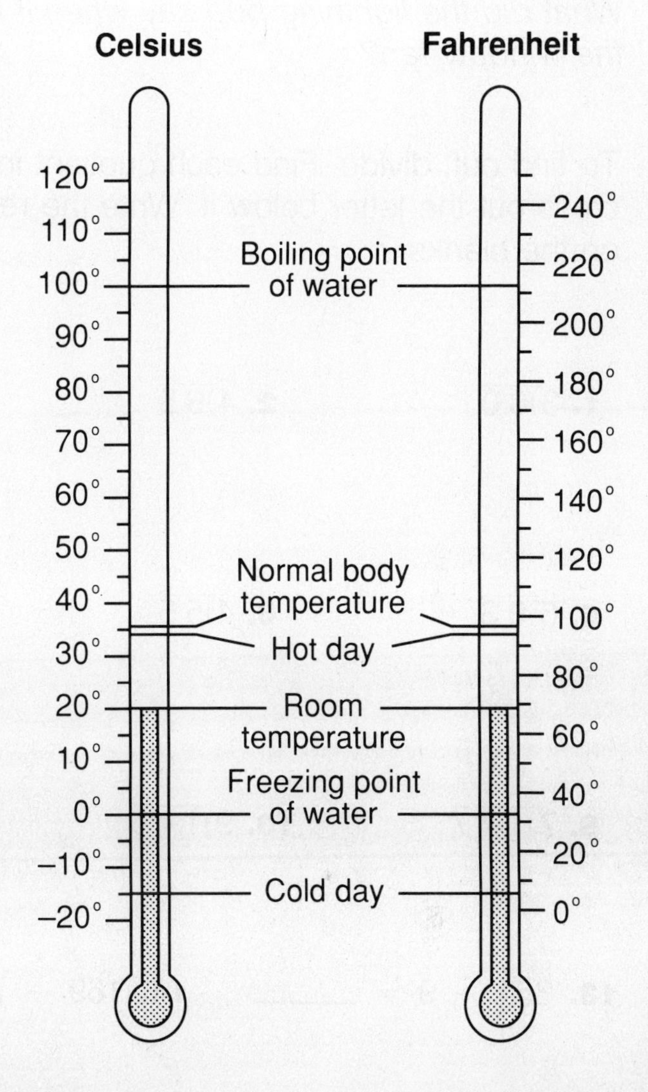

1. Hot casserole 170° C 17°C

2. Winter day in Detroit 1°C 35°C

3. Daytime in the desert 115°F 200°F

4. Summer day in Dallas 30°C 90°C

5. Warm shower 85°F 200°F

6. Lake water in summer 25°F 80°F

7. Boiling water 100°C 212°C

Find how many degrees the mark represents and give the temperature.

8. 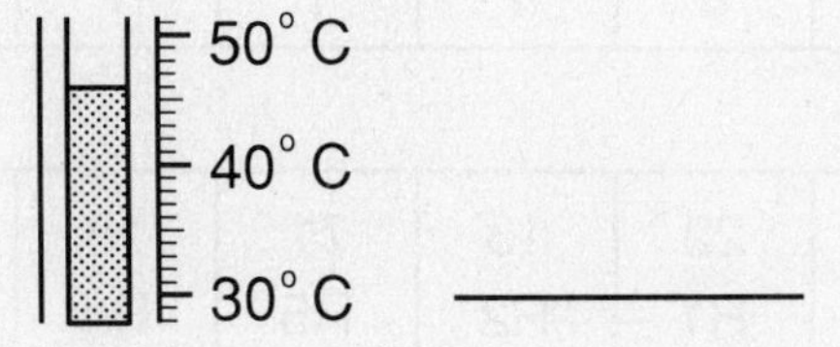

____

9. 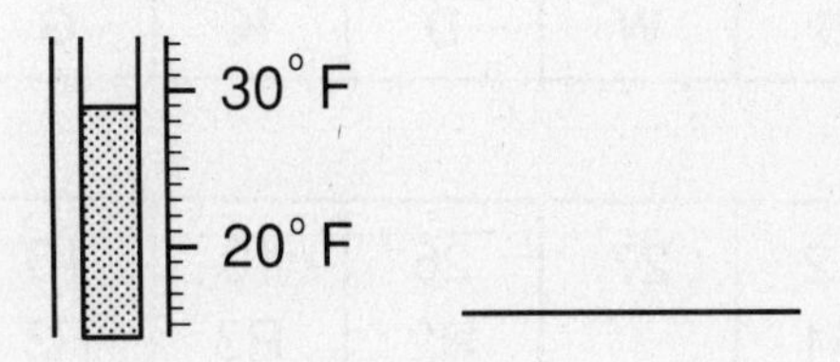

____

10. 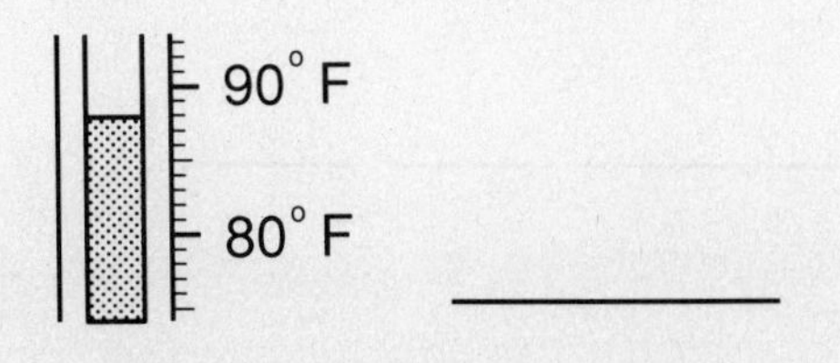

____

11.
40° C
30° C
20° C
____

NAME

SHARPEN YOUR SKILLS

Dividing Whole Numbers

What did the lightning bug say when it backed into the window fan?

To find out, divide. Find each quotient in the boxes and cross out the letter below it. Write the remaining letters on the blanks.

1. $3\overline{)80}$ **2.** $4\overline{)65}$ **3.** $6\overline{)80}$ **4.** $9\overline{)99}$

5. $8\overline{)93}$ **6.** $4\overline{)55}$ **7.** $7\overline{)83}$ **8.** $2\overline{)25}$

9. $7\overline{)177}$ **10.** $9\overline{)110}$ **11.** $7\overline{)155}$ **12.** $4\overline{)144}$

13. $354 \div 9 =$ ______ **14.** $169 \div 7 =$ ______ **15.** $431 \div 6 =$ ______

16 R1	25 R2	26 R4	39 R3	11 R6	19 R3	11	23 R7	15 R4	12 R2	11 R5	25 R1
B	W	D	K	G	E	F	L	I	Y	T	G

12 R1	27	26 R2	13 R3	16 R3	36	16	22 R1	13 R2	71 R5	18 R4	24 R1
A	H	O	J	T	R	E	M	S	C	D	N

I'm ____ ____ ____ ____ ____ ____ ____ ____ ____ **!**

NAME

SHARPEN YOUR SKILLS

Try and Check

The Great Mantini predicts future events. Help him predict some events. Solve each problem by trying and checking combinations from the lists below. Write the numbers of the combinations after the events.

1. This event will happen in 2050. The product of its two numbers is 18. The sum is 11. What is the event?

 ____

 ____

2. This event will happen in 2100. The product of its two numbers is 48. The sum is 16. What is the event?

 ____

 ____

3. This event will happen in 2001. The sum of its two numbers is 28. The product is 75. What is the event?

 ____

 ____

 ____

4. This event will happen in 2500. The sum of its two numbers is 17. The product is 70. What is the event?

 ____

 ____

 ____

25 Space beings discovered
2 Space colony is built
4 Gold is found
10 Spaceball

7 played in Asteroid Belt
12 on Mars
3 living under the ocean
9 on the Moon

NAME

SHARPEN YOUR SKILLS

Mental Math and Estimating Quotients

Estimate each quotient using compatible numbers. Tell what numbers you used. Then find each quotient.

1. 49 ÷ 8 **2.** 61 ÷ 10 **3.** 212 ÷ 4

4. 124 ÷ 5 **5.** 488 ÷ 8 **6.** 243 ÷ 3

7. 549 ÷ 9 **8.** 125 ÷ 6 **9.** 292 ÷ 6

10. 151 ÷ 5 **11.** 539 ÷ 9 **12.** 145 ÷ 7

13. 447 ÷ 9 **14.** 415 ÷ 4 **15.** 709 ÷ 8

16. 215 ÷ 3 **17.** 297 ÷ 6 **18.** 167 ÷ 4

Use after pages 126–127.

NAME

SHARPEN YOUR SKILLS

Dividing Larger Numbers

Find each quotient. Match each letter to its answer in the blanks below.

1. 2)790 A   **2.** 7)943 C   **3.** 8)908 E   **4.** 3)522 D

5. 4)649 N   **6.** 5)983 O   **7.** 2)1,758 S   **8.** 7)830 V

9. 3)804 R   **10.** 6)698 U   **11.** 5)765 G   **12.** 4)973 R

13. 9)1,648 E   **14.** 7)3,885 T   **15.** 5)2,322 N   **16.** 4)2,492 E

Jon and Tim wear trench coats to bed because they want to be

___ 116 R2 ___ 162 R1 ___ 174 ___ 113 R4 ___ 243 R1 ___ 134 R5 ___ 196 R3 ___ 118 R4 ___ 623 ___ 268

___ 395 ___ 153 ___ 183 R1 ___ 464 R2 ___ 555 ___ 879 when they grow up.

NAME

SHARPEN YOUR SKILLS

Zeros in the Quotient

Find each quotient.

1. $5\overline{)549}$ **2.** $6\overline{)722}$ **3.** $2\overline{)601}$ **4.** $3\overline{)2,132}$

5. $8\overline{)4,068}$ **6.** $9\overline{)3,605}$ **7.** $4\overline{)2,883}$ **8.** $7\overline{)4,201}$

9. $3\overline{)323}$ **10.** $8\overline{)5,619}$ **11.** $5\overline{)1,502}$ **12.** $9\overline{)8,556}$

13. $4\overline{)817}$ **14.** $9\overline{)3,788}$ **15.** $8\overline{)2,006}$ **16.** $5\overline{)4,536}$

Mixed Practice **Remember** to watch the signs.

17. 450 ÷ 9 ______ **18.** 11 × 49 ______ **19.** 3,561 − 476 ______ **20.** 568 + 239 ______

21. 3 × 5,611 ______ **22.** 932 − 454 ______ **23.** 324 ÷ 5 ______ **24.** 2,343 + 856 ______

25. 62 + 578 ______ **26.** 692 ÷ 8 ______ **27.** 4,982 − 1,698 ______ **28.** 8 × 108 ______

Use after pages 132–133.

NAME

SHARPEN YOUR SKILLS

Finding Averages

The Coyotes basketball team played 20 games last season.

1. Find their average score for each 5-game stretch. Use paper and pencil or a calculator.

5-game stretch	Scores					Average
First 5 games	18	16	23	26	22	
Second 5 games	24	29	30	20	27	
Third 5 games	23	21	34	33	24	
Fourth 5 games	26	25	27	35	37	

2. Find the average number of points for each player's 5 best games.

	Points					Average
Barb	7	6	9	10	8	
Beth	4	10	7	5	4	
Cathy	10	9	6	8	7	
Cindy	13	6	9	8	9	
Dan	8	10	11	7	9	
Joel	14	9	12	9	11	
Rick	12	10	8	7	13	
Tom	11	15	9	17	8	

3. Find the average number of foul shots attempted by each player.

	Foul shots				Average
Barb	16	23	19	14	
Beth	18	16	15	15	
Cathy	18	24	26	24	
Cindy	31	29	27	25	
Dan	23	29	25	23	
Joel	27	35	28	26	
Rick	35	32	29	32	
Tom	28	33	31	32	

Use after pages 134–135.

SHARPEN YOUR SKILLS

Write an Equation

Write an equation. Then find the answer.

1. Kate is reading a biography of Thomas Jefferson. It has 1,252 pages. Kate has read 786 pages. How many more pages does she have to read?

____

____

2. At the beginning of the school year, 19 boxes of books were unpacked. Each box contained 24 books. How many books were unpacked?

____

____

3. The 14 students in Mr. Ruisi's class read a total of 2,520 pages this week. What is the average number of pages each student read?

____

____

4. In one month, Adam read 609 pages, Bill read 782 pages, and Andy read 1,259 pages. How many pages did the boys read that month?

____

____

5. The school library has three sections of books. There are 1,301 books in one section, 1,094 in another, and 897 in the third. How many books are in the three sections?

____

____

6. In Mrs. Heller's class, 9 students each read the same number of books during the year. Altogether they read 261 books. How many books did each student read?

____

____

NAME ____________________

SHARPEN YOUR SKILLS

Missing Factors

The names of great sports heroes from the past are listed below. Match each name to the correct sport.

To check your answer, find n for each exercise. Match the answer with the exercise number.

1. Ben Hogan	baseball	$n \times 14 = 112$	$n = 8$
2. Maureen Connolly	horse racing	$n \times 26 = 78$	______
3. Willie Shoemaker	track	$n \times 67 = 603$	______
4. Joe Louis	golf	$99 \times n = 99$	______
5. Florence Chadwick	ice skating	$58 \times n = 580$	______
6. Red Grange	football	$20 \times n = 120$	______
7. Bob Cousy	swimming	$325 = 65 \times n$	______
8. Babe Ruth	tennis	$n \times 92 = 184$	______
9. Jesse Owens	boxing	$160 = n \times 40$	______
10. Dick Button	basketball	$33 \times n = 231$	______

NAME

SHARPEN YOUR SKILLS

Mental Math for Division

Use mental math to find each quotient.

1. 400 ÷ 20 = ______ **2.** 160 ÷ 10 = ______ **3.** 280 ÷ 40 = ______

4. 360 ÷ 60 = ______ **5.** 320 ÷ 40 = ______ **6.** 270 ÷ 30 = ______

7. 450 ÷ 50 = ______ **8.** 900 ÷ 30 = ______ **9.** 210 ÷ 70 = ______

10. 5,000 ÷ 10 = ____ **11.** 1,600 ÷ 20 = ____ **12.** 2,400 ÷ 60 = ____

13. 1,800 ÷ 20 = ____ **14.** 3,600 ÷ 30 = ____ **15.** 900 ÷ 10 = ______

16. 7,200 ÷ 80 = ____ **17.** 810 ÷ 90 = ______ **18.** 4,200 ÷ 60 = ____

19. 6,400 ÷ 80 = ____ **20.** 480 ÷ 60 = ______ **21.** 300 ÷ 10 = ______

22. 4,040 ÷ 40 = ____ **23.** 50,500 ÷ 50 = ____ **24.** 1,790 ÷ 10 = ____

25. 540 ÷ 90 = ______ **26.** 3,200 ÷ 80 = ____ **27.** 5,600 ÷ 70 = ____

28. 4,000 ÷ 50 = ____ **29.** 630 ÷ 70 = ______ **30.** 7,200 ÷ 90 = ____

Use after pages 150–151.

NAME

SHARPEN YOUR SKILLS

Estimating Quotients

Estimate each quotient. Tell what compatible numbers you used.

1. 184 ÷ 22 = ____ ________ ________

2. 319 ÷ 81 = ____ ________ ________

3. 724 ÷ 94 = ____ ________ ________

4. 636 ÷ 72 = ____ ________ ________

5. 404 ÷ 43 = ____ ________ ________

6. 454 ÷ 56 = ____ ________ ________

7. 876 ÷ 89 = ____ ________ ________

8. 257 ÷ 52 = ____ ________ ________

9. 919 ÷ 12 = ____ ________ ________

10. 3,221 ÷ 82 = ____ ________ ________

11. 7,195 ÷ 79 = ____ ________ ________

12. 4,501 ÷ 88 = ____ ________ ________

13. 8,167 ÷ 23 = ____ ________ ________

14. 5,679 ÷ 84 = ____ ________ ________

15. 6,351 ÷ 74 = ____ ________ ________

16. 2,998 ÷ 62 = ____ ________ ________

17. 1,784 ÷ 39 = ____ ________ ________

18. 9,111 ÷ 28 = ____ ________ ________

19. 7,689 ÷ 72 = ____ ________ ________

20. 3,976 ÷ 81 = ____ ________ ________

21. 5,425 ÷ 89 = ____ ________ ________

SHARPEN YOUR SKILLS

One-Digit Quotients

Tell whether you would use paper and pencil or mental math. Then find each quotient.

1. $40\overline{)82}$ ____

2. $23\overline{)52}$ ____

3. $47\overline{)52}$ ____

4. $13\overline{)29}$ ____

5. $52\overline{)479}$ ____

6. $79\overline{)328}$ ____

7. $22\overline{)76}$ ____

8. $63\overline{)315}$ ____

9. $84\overline{)513}$ ____

10. $56\overline{)191}$ ____

11. $47\overline{)94}$ ____

12. $41\overline{)386}$ ____

13. $90\overline{)724}$ ____

14. $98\overline{)217}$ ____

15. $67\overline{)290}$ ____

16. $86\overline{)192}$ ____

Connect the dots in the order of the answers to find out why the cat is smiling.

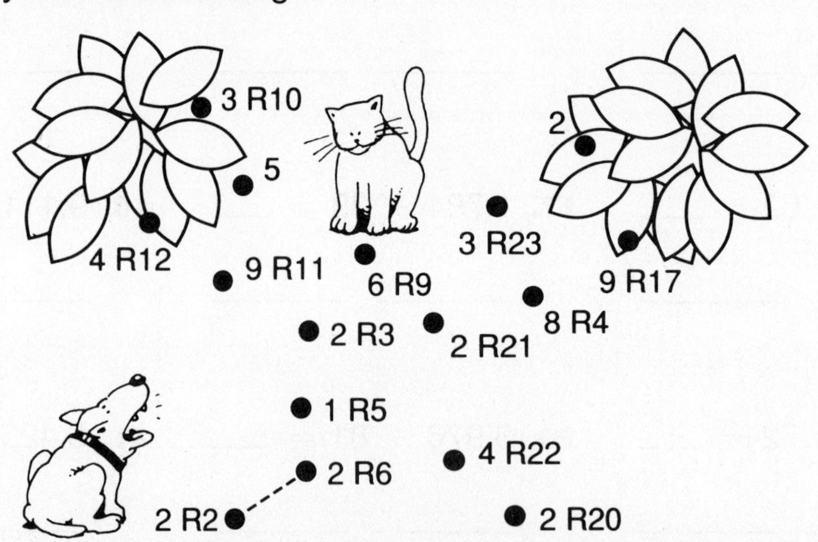

NAME

SHARPEN YOUR SKILLS

Adjusting the Quotient

Divide. Show each answer as a quotient with a whole number remainder (if there is one) or as a decimal with one digit after the decimal point.

1. $79\overline{)147}$ **2.** $39\overline{)78}$ **3.** $35\overline{)239}$

4. $26\overline{)125}$ **5.** $84\overline{)766}$ **6.** $81\overline{)555}$

7. $42\overline{)252}$ **8.** $23\overline{)114}$ **9.** $21\overline{)143}$

10. $32\overline{)121}$ **11.** $36\overline{)90}$ **12.** $63\overline{)118}$

13. $79\overline{)716}$ **14.** $46\overline{)328}$ **15.** $96\overline{)827}$

Cross out each box that has an answer. Some answers are not used. You will find a message in the remaining letters.

Y	M	B	A	T	Z	S	T	U	H
1 R68	2 R17	2 R18	6 R30	6 R28	6 R29	9 R10	6 R17	7 R6	3 R80
1.8	2.8	2.5	6.9	6.3	6.8	9.1	6.8	7.1	3.8

I	S	S	K	T	F	A	U	R	N
1 R70	9 R6	8 R59	6	9 R5	3 R75	3 R25	2 R16	4 R21	1 R2
1.7	9.1	8.6	6.0	9.0	3.8	3.7	2.7	4.8	1.0

NAME

SHARPEN YOUR SKILLS

Using Rounded Divisors

Use paper and pencil or a calculator. Show each answer as a quotient with a whole number remainder (if there is one) or as a decimal with one digit after the decimal point.

1. $46\overline{)218}$ **2.** $68\overline{)321}$ **3.** $66\overline{)594}$ **4.** $87\overline{)337}$

5. $78\overline{)502}$ **6.** $28\overline{)140}$ **7.** $39\overline{)284}$ **8.** $59\overline{)377}$

9. $17\overline{)155}$ **10.** $18\overline{)161}$ **11.** $78\overline{)290}$ **12.** $88\overline{)648}$

Mixed Practice Tell whether you would use paper and pencil, a calculator, or mental math. Then find each answer.

13. 228 × 38 = ______ **14.** 359 + 321 = ______ **15.** 4,600 − 300 = ______

16. 173 ÷ 28 = ______ **17.** 360 ÷ 60 = ______ **18.** 215 × 17 = ______

19. 410 + 210 = ______ **20.** 971 − 629 = ______ **21.** 762 ÷ 89 = ______

Use after pages 158–159.

NAME

SHARPEN YOUR SKILLS

Two-Digit Quotients

Use paper and pencil or a calculator. Show each answer as a quotient with a whole number remainder (if there is one) or as a decimal with one digit after the decimal point.

1. $20\overline{)675}$ **2.** $74\overline{)845}$ **3.** $53\overline{)810}$ **4.** $21\overline{)567}$

5. $38\overline{)930}$ **6.** $13\overline{)276}$ **7.** $43\overline{)933}$ **8.** $62\overline{)930}$

9. 1,134 ÷ 21 = ______ **10.** 5,690 ÷ 67 = ______

11. 3,522 ÷ 75 = ______ **12.** 2,980 ÷ 94 = ______

Solve the problem.

13. The Commans family flew a distance of 736 miles in their airplane. They used 92 gallons of gasoline. How many miles did they fly per gallon of gasoline? ______

SHARPEN YOUR SKILLS

Interpret the Remainder

Solve each problem. Pay close attention to the remainders.

1. Rick has $8 to spend at the school carnival. He bought as many books of tickets for $3 a book as he could. How much money did Rick have left?

____

2. Only 22 people can be seated in the Funny Flicks Theater. How many times must a film be shown so that 300 people can see it?

____

3. The poster committee worked for 26 hours on signs. They spent 3 hours on each sign. How many signs did they complete?

____

4. Kelly had 103 streamers. Each booth was decorated with 12 streamers. How many streamers were left over?

____

5. Billy used 5,286 centimeters of crepe paper to make streamers 94 centimeters long. How many streamers did Billy make?

____

6. One package of crepe paper will cover 3 tables. How many packages would be needed to cover 40 tables?

____

7. Masako needs 700 balloons for the booths. Balloons come in packages of 36 each. How many packages should Masako buy?

____

8. Booth workers wear smocks made of 88 centimeters of material. How many smocks could be made from 800 centimeters of material?

____

Use after pages 164–165.

NAME

SHARPEN YOUR SKILLS

Three-Digit Quotients

Use paper and pencil or a calculator. Show each answer as a quotient with a whole number remainder (if there is one) or as a decmial with one digit after the decimal point.

1. $40\overline{)5{,}815}$

2. $82\overline{)9{,}524}$

3. $28\overline{)4{,}312}$

4. $53\overline{)7{,}093}$

5. $71\overline{)9{,}099}$

6. $32\overline{)8{,}548}$

7. $9{,}201 \div 64 =$ ________________

8. $5{,}633 \div 43 =$ ________________

Solve each problem.

9. Darren spoke for 9,776 minutes on the telephone last year. What was the average number of minutes that Darren spoke on the phone per week? (52 weeks = 1 year)

____

10. There are 2,688 names in the school telephone directory, listed on 14 pages. How many names are listed on each page?

____

NAME

SHARPEN YOUR SKILLS

Zeros in the Quotient

Use paper and pencil or a calculator. Show each answer as a quotient with a whole number remainder (if there is one) or as a decimal with one digit after the decimal point.

1. $61\overline{)2{,}476}$

2. $22\overline{)3{,}315}$

3. $83\overline{)8{,}353}$

4. $28\overline{)8{,}417}$

5. $94\overline{)5{,}701}$

6. $33\overline{)6{,}917}$

7. 1,835 ÷ 36 = ______________

8. 9,020 ÷ 82 = ______________

Find each correct answer in the number chart. Cross out the letter below the correct quotient. The letters that remain will tell you who is a special person.

300 R17 300.6	150 R15 150.6	209 R20 209.6	100 R53 100.6	107 R31 107.6	50 R35 50.9	40 R36 40.5	60 R61 60.6	110 110.0	230 230.0
A	**P**	**N**	**D**	**U**	**O**	**F**	**I**	**K**	**R**

NAME

SHARPEN YOUR SKILLS

Choose an Operation

Tell whether to add, subtract, multiply, or divide.
Then solve the problem.

1. In a dictionary 15 pages are 1 millimeter thick. What is the thickness of 1,050 pages of this dictionary?

____

2. A book contains 12 sections with 64 pages in each section. What is the total number of pages in the book?

____

3. A book company received an order for 5,400 books. If there are 36 books in each box, how many boxes are needed to fill the order?

____

4. A paperback book costs $3.98. The same book with a hard cover costs $8.95. What is the difference between the two prices?

____

5. The charge for shipping one box of books is $5.29. How much does it cost to ship 38 boxes of books?

____

6. The library has 8 sets of encyclopedias totaling 256 books. What is the average number of books in each set?

____

7. The school library spent $651 for math books, $823 for science books, and $1,342 for language books. How much money was spent in all?

____

SHARPEN YOUR SKILLS

Investigating Solid Shapes

Tell whether each object is like a cube, like a rectangular prism, like both a cube and a rectangular prism, or like neither.

1.  ________________

2. 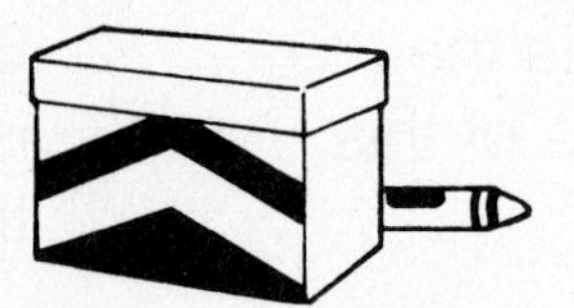 ________________

3.  ________________

4.  ________________

5. 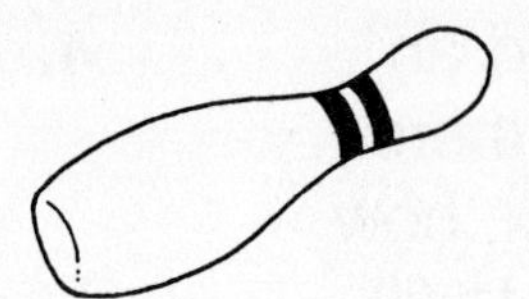 ________________

6. 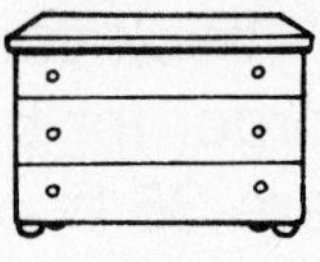 ________________

7.  ________________

8. 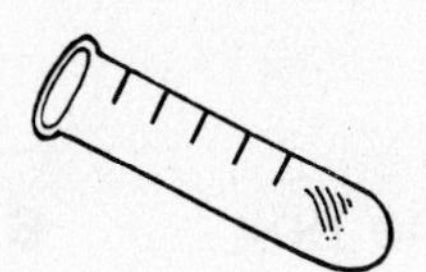 ________________

9. 

____

10.  ________________

11. 

____

12.  ________________

Answer the question.

13. What is the difference between a cube and a rectangular prism?

____

____

NAME

SHARPEN YOUR SKILLS

Finding Volume by Counting Cubes

Each small cube in Exercises 1–6 represents 1 cubic centimeter. Find the volume of each figure.
Remember to write the unit of measure.

1. 

2. 

3. 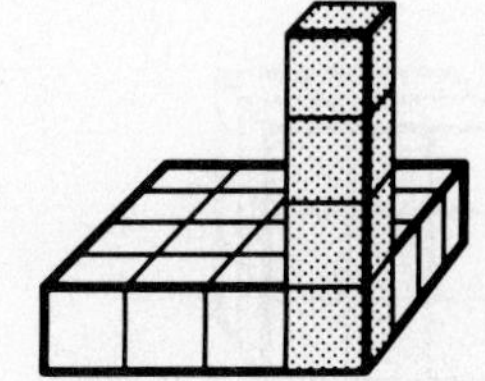

4. 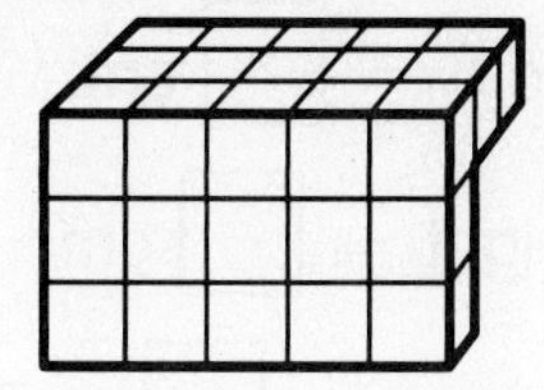

5. 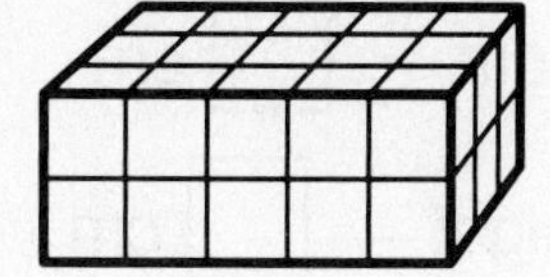

6. 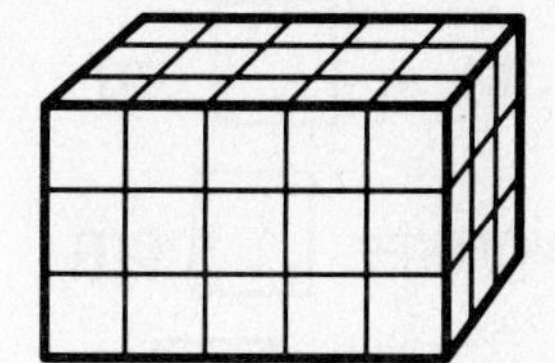

Each cube in Exercises 7–12 represents 1 cubic inch.
Find the volume of each figure.
Remember to write the unit of measure.

7. 

8. 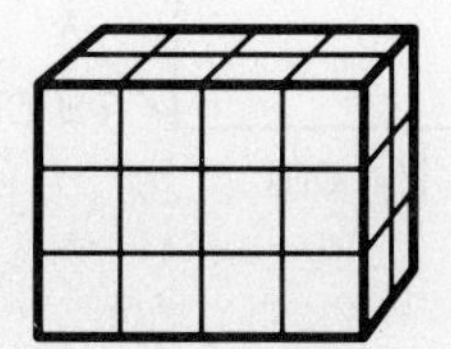

9. 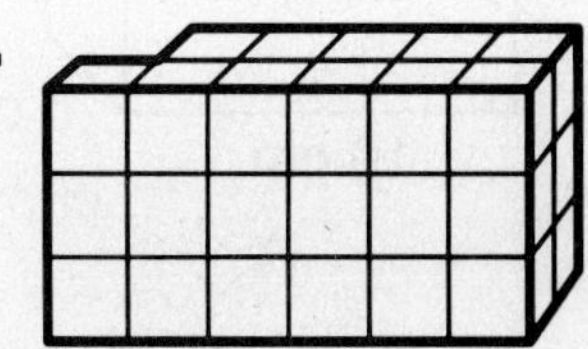

10. 

11. 

12. 

Use after pages 184–185.

SHARPEN YOUR SKILLS

Volume

Fill in the missing numbers for Exercises 1–3.
Then find the volume. Each cube is 1 cubic centimeter.

1.

Length = ☐ cm

Width = ☐ cm

Height = ☐ cm

Volume = ☐ cm^3

2.

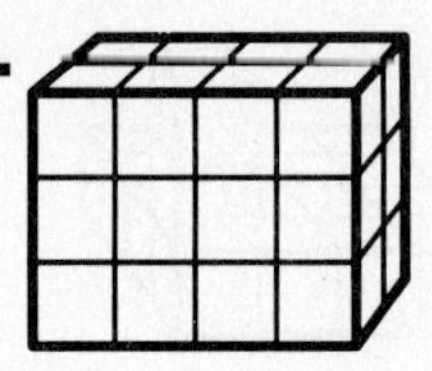

Length = ☐ cm

Width = ☐ cm

Height = ☐ cm

Volume = ☐ cm^3

3.

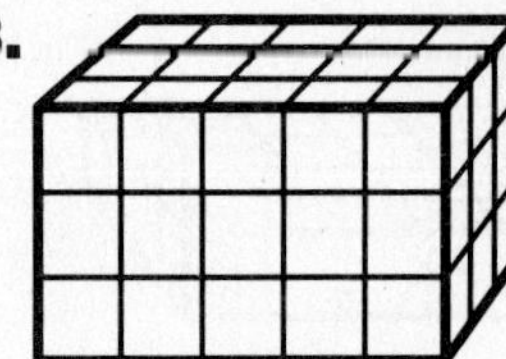

Length = ☐ cm

Width = ☐ cm

Height = ☐ cm

Volume = ☐ cm^3

For Exercises 4–9, write *P, M,* or *C* for whether you would use paper and pencil, mental math, or a calculator to find the volume of each figure. Then find the volume.

4.

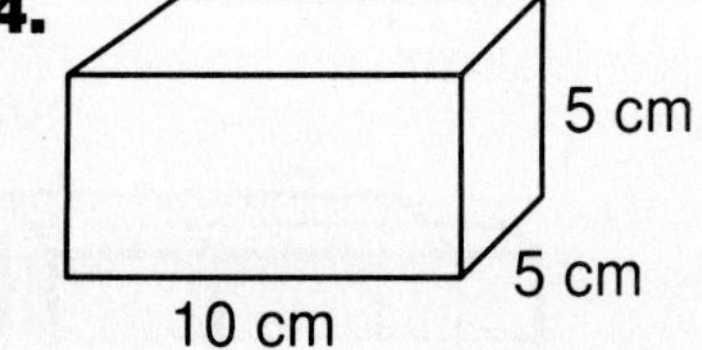

____

5.

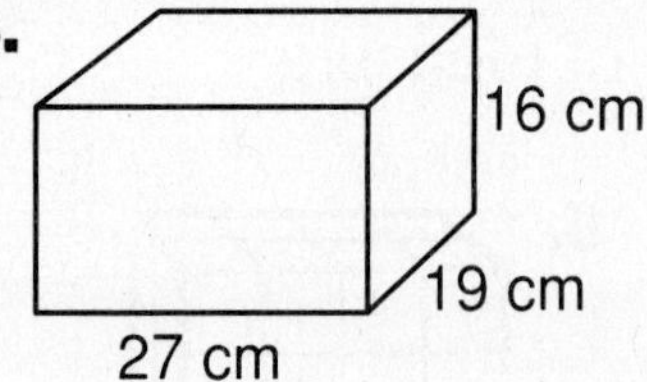

____

6.

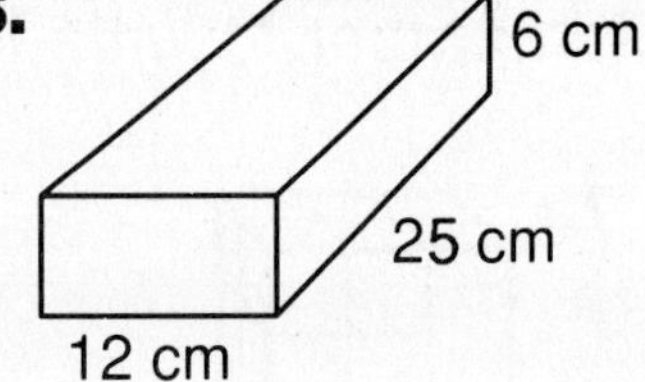

____

7.

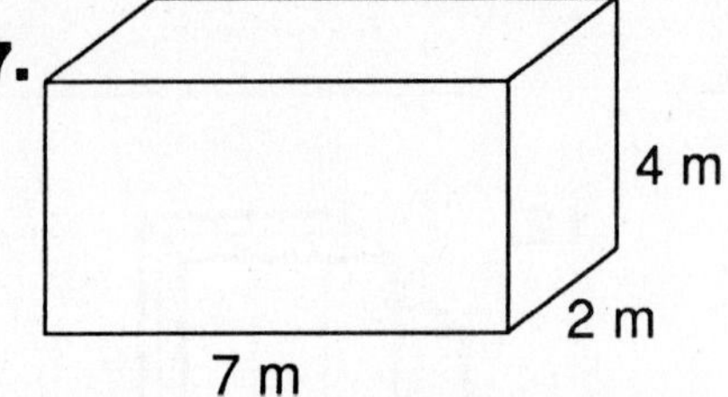

____

8.

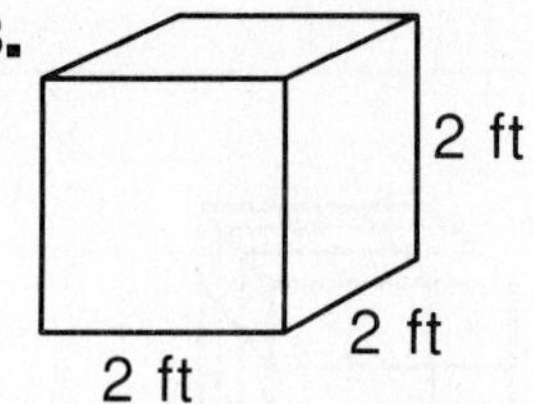

____

9.

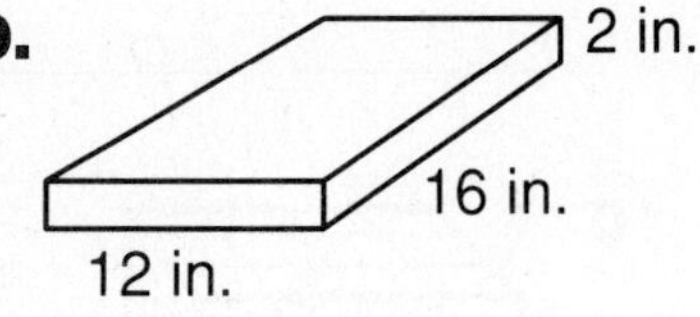

____

Critical Thinking Which problems are given in customary units of measurement? ______________

NAME

SHARPEN YOUR SKILLS

Customary Units of Capacity

Estimation Choose the most sensible measure.

1. Bottle of perfume

$\frac{1}{2}$ c $\frac{1}{2}$ qt $\frac{1}{2}$ gal

2. Milk bottle

2 c 2 pt 2 gal

3. Teapot

8 c 8 qt 8 gal

4. Washing machine

25 pt 25 qt 25 gal

5. Thermos bottle

1 c 1 qt 1 gal

6. Gasoline container

5 c 5 qt 5 gal

Write each measure in pints.

7. 5 gal = __________ pt

8. 10 qt = __________ pt

9. 16 c = __________ pt

10. 4 c = __________ pt

Write each measure in quarts.

11. 16 gal = __________

12. 8 pt = __________

Solve the problem.

13. Gloria wants to buy 4 pints of strawberries. They are sold in pints for $0.79 and in quarts for $1.50. Should Gloria buy the strawberries in pint or quart containers? What is the difference in price?

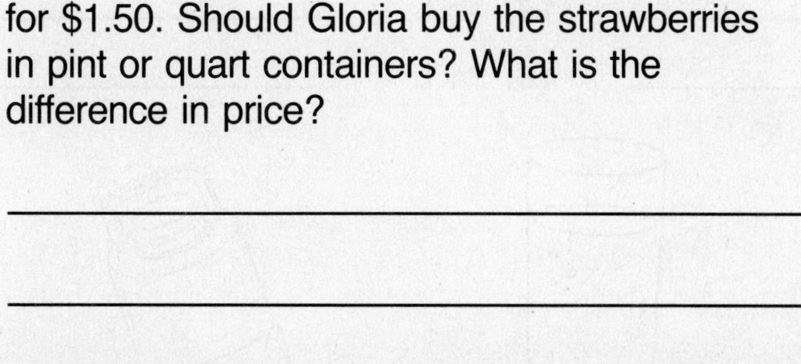

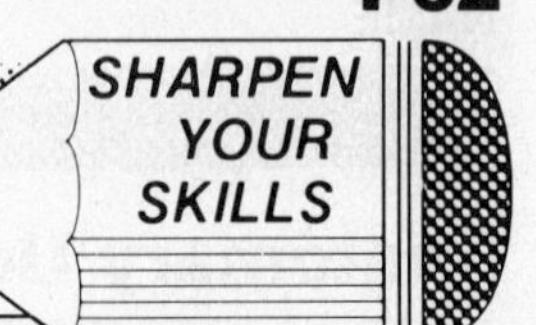

Metric Units of Capacity

Estimation Choose the most sensible answer.

1. Soup cup

22 mL 220 mL 22,000 mL

2. Sink

250 L 2,500 L 25 L

3. Bottle of dish detergent

450 mL 45 mL 4,500 mL

4. Thermos bottle

850 mL 8,500 mL 85 mL

5. Juice can

175 mL 1,750 mL 17,500 mL

6. Teapot

10 L 1 L 100 L

7. Eyedropper

50 mL 5 mL 500 mL

8. Glass of water

235 mL 2,350 mL 23,500 mL

Use estimation to solve each problem.

9. If a cup holds about 140 mL, about how many full cups of orange juice can Hillary pour from a 1-liter container?

____

10. About how much orange juice will be left in the container?

____

Write each measure in milliliters.

11. 5 L = ______________

12. 48 L = ______________

13. 256 L = ______________

14. 15 L = ______________

15. 70 L = ______________

16. 94 L = ______________

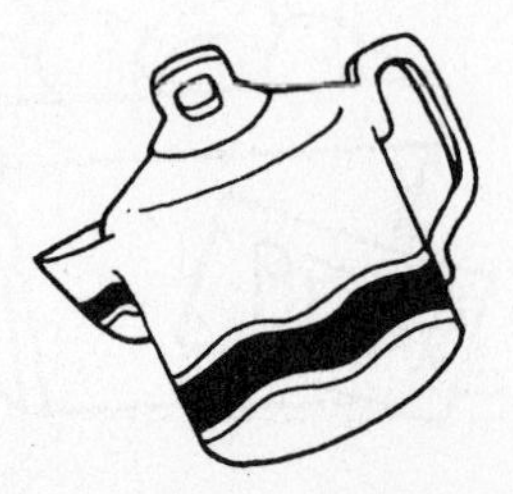

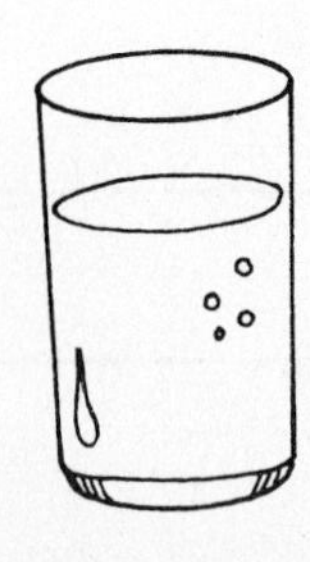

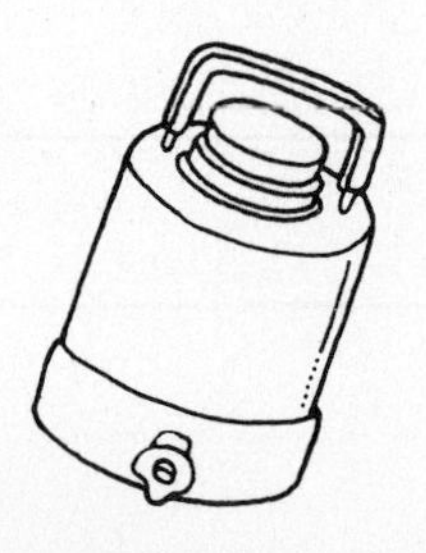

NAME

SHARPEN YOUR SKILLS

Customary Units of Weight

Write each measure in ounces.

1. 5 lb = ________ **2.** 10 lb = ________ **3.** 14 lb = ________

4. 7 lb 6 oz = ________ **5.** 61 lb 15 oz = ________

Write each measure in pounds.

6. 2 T = ________ **7.** 192 oz = ________

8. 256 oz = ________ **9.** 12 T = ________

Write each measure in pounds and ounces.

10. 30 oz = ________ **11.** 60 oz = ________

12. 85 oz = ________ **13.** 100 oz = ________

Solve each problem.

14. A box of detergent weighs 6 pounds. What is the total weight of 4 boxes?

____

15. Tony collected 40 pounds of newspaper on the first day of the newspaper collection drive. If he collects the same amount on each day of the 9-day drive, how many pounds will he collect?

____

NAME

SHARPEN YOUR SKILLS

Metric Units of Mass

Circle the more sensible measure.

1. Dime

3g   3 kg

2. Adult

77 g   77 kg

3. Brick

1 g   1 kg

4. Cat

4 g   4 kg

5. Toothpaste tube

150 g   150 kg

6. Car

1,500 g   1,500 kg

Write each measure in grams.

7. 7 kg = ____________________

8. 17 kg = ____________________

9. 40 kg = ____________________

10. 88 kg = ____________________

11. 10 kg = ____________________

12. 150 kg = ____________________

Write each measure in kilograms.

13. 6,000 g = ____________________

14. 37,000 g = ____________________

15. 10,000 g = ____________________

16. 79,000 g = ____________________

17. 14,000 g = ____________________

18. 11,000 g = ____________________

Solve each problem.

19. A box of cereal weighs 444 grams. How many 37-gram servings are in the box?

____

20. How many kilograms are there in 21 bags of flour that each weigh 5,000 grams?

____

NAME

SHARPEN YOUR SKILLS

Try and Check

Eric is boxing an order placed by a restaurant. The restaurant doesn't want any box to weigh more than 40 pounds.

Restaurant Order	Number	Weight
sack of potatoes	3	20 lb
bag of onions	4	5 lb
can of juice	8	6 oz
bag of flour	1	15 lb
tomato paste	8	4 oz

Find the total weight of the items in pounds.

1. potatoes ________

2. juice ________

3. tomato paste ________

4. onions ________

5. Identify what Eric should put in each box. Write the items in the boxes.

Box 1	Box 2	Box 3

6. If Eric uses four boxes, what is the least amount of weight he can put in the fourth box? Explain.

____

____

7. Can Eric put a bag of potatoes, the bag of flour, a bag of onions, and 3 cans of juice in one box? Explain.

____

8. Can Eric pack 3 boxes so that they all contain equal weight? Explain.

____

____

Use after pages 198–199.

NAME

SHARPEN YOUR SKILLS

Computing with Customary Measures

Mental Math Use mental math for Exercises 1–4. Find the missing number.

1. 5 gal 3 qt = 4 gal ☐ qt

2. 8 ft 13 in. = 9 ft ☐ in.

3. 10 lb 22 oz = 11 lb ☐ oz

4. 5 qt 2 c = 4 qt ☐ c

Add or subtract.

5. 23 lb 14 oz − 6 lb 6 oz

6. 5 qt 1 c + 1 qt 2 c

7. 6 yd + 8 yd 2 ft

8. 33 lb 13 oz + 9 lb 14 oz

9. 48 gal 3 qt + 26 gal 2 qt

10. 83 ft 6 in. − 6 ft 10 in.

11. 4 ft 16 in. − 1 ft 19 in.

12. 10 ft 4 in. + 5 ft 6 in.

13. 2 ft 11 in. + 9 ft 9 in.

Multiply or divide.

14. 2 × 2 lb 4 oz ______

15. 3 × 9 qt 1 pt ______

16. 5 × 16 ft 12 in. ______

17. 8 ft 8 in. ÷ 4 ______

18. 3 qt 2 c ÷ 7 ______

19. 6 ft 15 in. ÷ 3 ______

SHARPEN YOUR SKILLS

Use Data from a Table

Use the table of prices for horizontal blinds to complete the exercises.

Prices for $\frac{1}{2}$-inch Made-to-Order Horizontal Blinds

WIDTH→ LENGTH ↓	16 to 23 in.	$23\frac{1}{8}$ to 26 in.	$26\frac{1}{8}$ to 29 in.	$29\frac{1}{8}$ to 32 in.	$32\frac{1}{8}$ to 36 in.	$36\frac{1}{8}$ to 40 in.	$40\frac{1}{8}$ to 44 in.	$44\frac{1}{8}$ to 48 in.
12 to 42 in.	40.00	44.00	48.00	51.00	58.00	63.00	67.00	71.00
$42\frac{1}{8}$ to 48 in.	44.00	48.00	52.00	56.00	64.00	68.00	73.00	78.00
$48\frac{1}{8}$ to 54 in.	47.00	52.00	56.00	61.00	68.00	73.00	78.00	85.00
$54\frac{1}{8}$ to 60 in.	51.00	55.00	60.00	65.00	73.00	78.00	85.00	91.00
$60\frac{1}{8}$ to 66 in.	56.00	60.00	66.00	71.00	78.00	84.00	91.00	97.00

When the length of the blinds being ordered is 55 inches, what would the price be if the width of the blinds is

1. 43 in.? ______ **2.** $26\frac{1}{2}$ in.? ______ **3.** 16 in.? ______ **4.** 35 in.? ______

When the width of the blinds being ordered is 35 inches, what would the price be if the length of the blinds is

5. 56 in.? ______ **6.** $60\frac{1}{2}$ in.? ______ **7.** 42 in.? ______ **8.** $59\frac{1}{2}$ in.? ______

When the blinds being ordered are 60 inches long, what widths could they be if the blinds cost

9. \$65.00? ______ **10.** \$51.00? ______ **11.** \$78.00? ______ **12.** \$55.00? ______

NAME

SHARPEN YOUR SKILLS

Tenths and Hundredths

Write a fraction and a decimal for each exercise.

1. 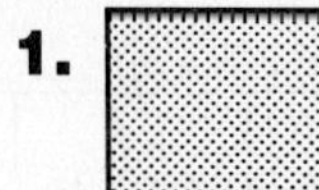 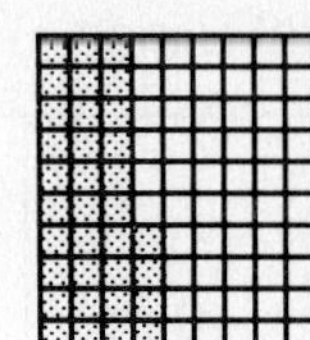

____

2. 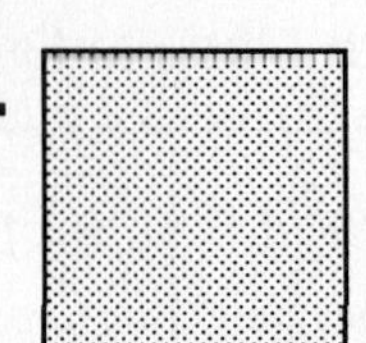 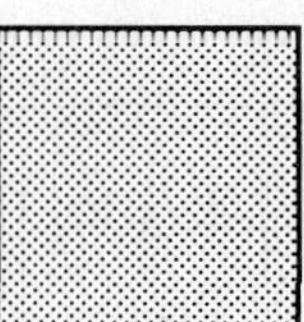 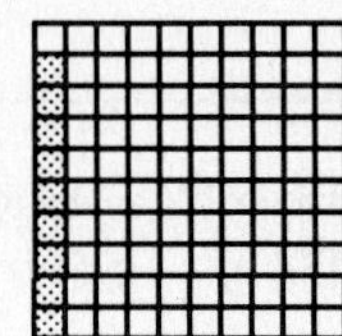

____

3. Thirty-three hundredths ____________

4. Twenty-nine and five hundredths ____________

5. Fourteen and nine tenths ____________

6. One and eleven hundredths ____________

7. One hundred and one tenth ____________

Write each decimal in words.

8. 68.7 ______________________________

9. 405.13 ______________________________

Write each price with a dollar sign and a decimal point.

10. A skateboard costs twenty-nine dollars and ninety-five cents.

____

11. A ten-speed bicycle costs one hundred nine dollars and ninety-nine cents.

____

NAME

SHARPEN YOUR SKILLS

Thousandths

Write each number in words. **Remember** to use "and" for decimals greater than 1.

1. 0.233 ______

2. 0.009 ______

3. 14.016 ______

4. 57.380 ______

5. 89.667 ______

Write a fraction and a decimal for each exercise.

6. Seven hundred two thousandths ______

7. Six and forty-three thousandths ______

8. Eight thousandths ______

9. Nine and fifty thousandths ______

10. Twenty-two thousand and one tenth ______

Write each number in words.

11. 0.015 ______

12. 30.002 ______

NAME ______ **P70**

SHARPEN YOUR SKILLS

Place Value

In each exercise, write the digit in the given place.
Then use the code below to find the hidden message.

1. 56.34	*tenths*	____		**2.** 231.56	*tenths*	____
3. 17.033	*hundredths*	____		**4.** 4.899	*thousandths*	____
5. 0.4	*tenths*	____		**6.** 58.019	*thousandths*	____
7. 4.365	*hundredths*	____		**8.** 5.738	*hundredths*	____
9. 4.8	*tenths*	____		**10.** 0.004	*thousandths*	____
11. 9.646	*tenths*	____		**12.** 14.257	*thousandths*	____
13. 15.462	*tenths*	____		**14.** 509.332	*hundredths*	____
15. 265.015	*hundredths*	____		**16.** 0.071	*thousandths*	____
17. 33.403	*hundredths*	____		**18.** 27.38	*tenths*	____
19. 25.75	*tenths*	____		**20.** 1.681	*tenths*	____
21. 31.052	*thousandths*	____		**22.** 178.56	*hundredths*	____
23. 0.806	*thousandths*	____		**24.** 78.96	*tenths*	____

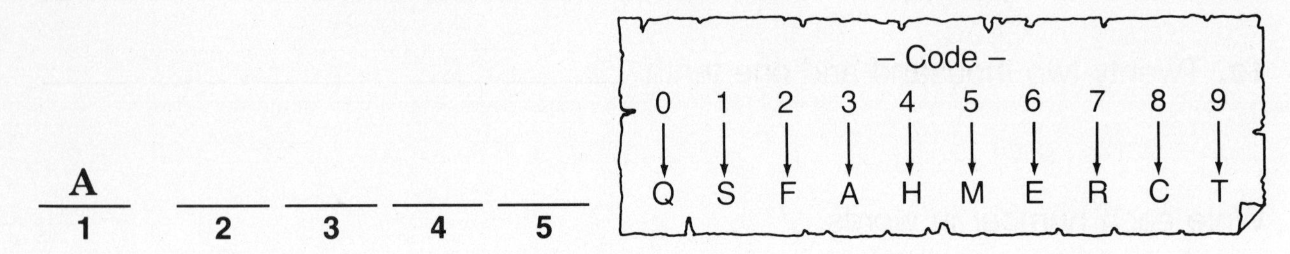

____ ____ ____ ____ ____ ____ ____   ____ ____ ____
6 7 8 9 10 11 12   13 14 15

____ ____ U ____ ____ ____   ____ ____ ____ ____ .
16 17   18 19 20   21 22 23 24

SHARPEN YOUR SKILLS

Comparing and Ordering Decimals

Compare the numbers. Use <, >, or =.
Draw a diagram if needed.

1. 0.4 ◯ 0.2 **2.** 0.3 ◯ 0.9 **3.** 0.10 ◯ 0.1

4. 0.27 ◯ 0.72 **5.** 0.50 ◯ 0.06 **6.** 44.50 ◯ 44.500

7. 8.24 ◯ 6.18 **8.** 0.70 ◯ 0.07 **9.** 30.52 ◯ 3.052

10. 9.35 ◯ 5.06 **11.** 0.2 ◯ 0.200 **12.** 27.014 ◯ 27.041

List the numbers in order from least to greatest. Use a number line if needed.

13. 13.4 13.9 13.1 ______________

14. 9.5 9.7 9.3 ______________

15. 0.563 0.568 0.564 ______________

16. 45.709 45.079 45.097 ______________

17. 58.88 58.99 58.77 ______________

18. 83.930 83.093 83.830 ______________

19. 75.099 75.009 75.101 ______________

Solve each problem.

20. The Arnold family drove 15.3 kilometers to visit the amusement park. The Banks family drove 15.9 kilometers. Which family drove farther?

____

21. The Hightower family drove 27.1 kilometers to visit the Triple Q Ranch. The Rivera family drove 26.9 kilometers. Which family drove farther?

____

SHARPEN YOUR SKILLS

Use Logical Reasoning

Solve each problem. **Remember** to make a chart to help you.

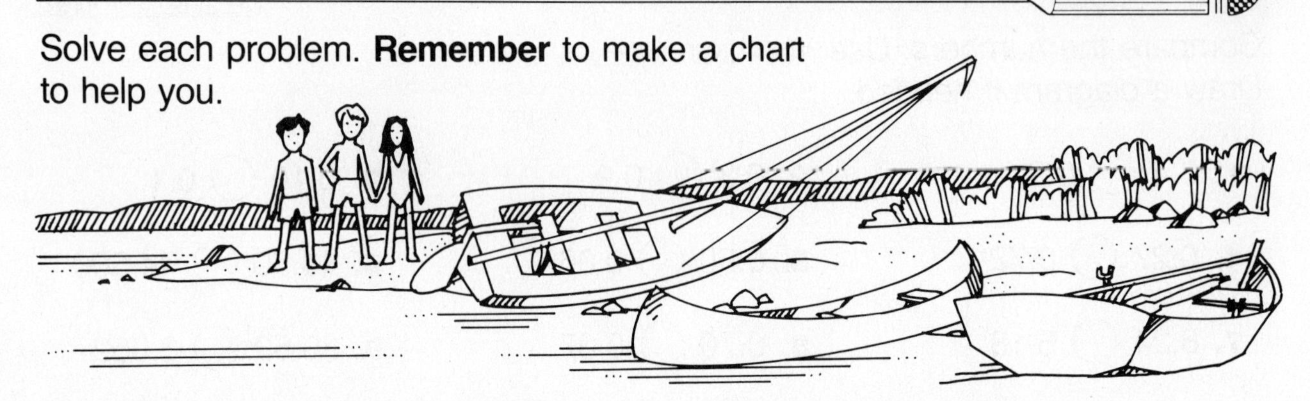

1. Frank, Ellen, and Juan each have a boat. One is a rowboat, one is a canoe, and one is a sailboat. One of the boys has the rowboat. The sailboat is parked next to Ellen's boat. Juan does not have a rowboat. Which boat does each person own?

____

____

2. Clark, Mary, Julie, and Joe each ate a different fruit: apple, banana, pear, and peach. Neither of the boys ate a peach. Julie enjoyed her apple. Clark wished he had eaten a pear. Which fruit did each person eat?

____

____

3. Rob, Francine, Billy, and Dee are in a track meet. Each person is in one event. The events are the high jump, 50-yard dash, long jump, and 100-yard dash. Dee's event requires a metal bar. Neither Billy nor Rob is in the long jump. Billy lives next door to the person who is doing the 100-yard dash. Who is running the 100-yard dash?

____

4. Patty, Mark, Jiro, and Olivia have a bicycle, a scooter, a skateboard, and roller skates. Neither Patty nor Olivia has a scooter. Neither Patty nor Mark has a skateboard or roller skates. The person with the skateboard is a good friend of Olivia. Which item does each person own?

____

____

____

NAME

SHARPEN YOUR SKILLS

Rounding Decimals

Round to the nearest whole number.

1. 3.6 ______ **2.** 4.1 ______ **3.** 5.7 ______ **4.** 35.4 ______ **5.** 16.9 ______

Round to the nearest tenth.

6. 4.38 ______ **7.** 18.09 ______ **8.** 6.505 ______ **9.** 7.89 ______ **10.** 0.66 ______

11. 11.32 ______ **12.** 2.063 ______ **13.** 84.48 ______ **14.** 3.62 ______ **15.** 14.91 ______

Round to the nearest hundredth.

16. 7.049 ______ **17.** 3.254 ______ **18.** 0.008 ______ **19.** 0.192 ______ **20.** 6.938 ______

Solve each problem.

21. Jose, Martin, and Greta ran a race. Jose had a time of 2.34 minutes. Martin ran the race in 2.43 minutes, and Greta ran it in 2.24 minutes. In what order did they finish?

____

22. In their second heat the times of the three racers were the following: Jose—2.29 minutes, Martin—2.40 minutes, and Greta—2.31 minutes. In what order did they finish the second heat?

____

NAME

SHARPEN YOUR SKILLS

Estimating Sums and Differences

Estimate each sum or difference.
Round to the nearest whole number.

1. $12.6 - 4.2$

2. $35.17 + 22.41$

3. $42.6 - 12.7$

4. $6.99 - 0.22$

5. $10.56 + 15.23$

6. $44.16 - 22.67$

7. $19.8 - 2.5$

8. $2.39 + 8.44$

9. $34.87 - 12.12$

10. $8.3 + 7.9$

Estimate each sum or difference.
Use front-end digits.

11. 15.3 + 4.9 ______

12. 44.3 − 31.8 ______

13. 9.7 + 76.9 ______

14. 65.362 − 39.243 ______

15. 14.32 + 63.76 ______

16. 55.032 − 3.981 ______

17. 62.10 − 19.54 ______

18. 13.6 + 21.4 ______

19. 5.04 − 3.62 ______

20. Compare your estimates for 4 of the exercises above with the exact answers.

Actual: ______ ______ ______ ______

Estimate: ______ ______ ______ ______

Too Much or Too Little Information

Solve each problem. If there is not enough information given to do so, write *too little information.*

Mrs. Dunn's class spent a week at a nature studies camp.

1. Each of the 20 students paid $100 for the trip. At the end of the week, $8.50 was returned to each student. What was the final cost for each student?

____

2. Before leaving, 10 students checked the supplies. Chris counted 30 dishes, and Tim counted 33 cups. How many more cups than dishes were packed?

____

3. The school bus traveled 256.3 miles to the camp. The bus returned to the school 9 hours later. What time did the bus leave the camp?

____

4. Eight students hiked on the forest trail, and 13 hiked on the river trail. The river trail was 3.6 miles long. How much longer was the forest trail?

____

5. Kim collected a piece of limestone that weighed 13 ounces. Luis collected a handful of fossil rocks that weighed 7 ounces. How much did all the rocks weigh?

____

6. The counselors told stories at bedtime. By the end of the week, Darnell had told 10 stories, Kendra had told 6, and Mike had told 2. How many more stories did Darnell tell than Kendra?

____

SHARPEN YOUR SKILLS

Adding Decimals

Add. **Remember** to estimate to be sure your answer makes sense. Each time an answer is given in the code below, write the letter for that exercise. You will see a well-known fact.

1. $0.4 + 0.5$ R

2. $0.8 + 0.5$ E

3. $0.8 + 2.9$ T

4. $7 + 25.5$ A

5. $6.6 + 58.5$ E

6. $75.21 + 4.79$ E

7. $9.04 + 94.07$ D

8. $33.4 + 26.86$ H

9. $62.1 + 14.7$ T

10. $10.66 + 14.92$ V

11. $5.827 + 2.19$ O

12. $5.064 + 8.379$ O

13. $47.483 + 2.10 + 18.65$ ______ A

14. $3.6 + 0.7 + 8.36$ ______ R

15. $82.6 + 41.8 + 7.4$ ______ U

16. $4.6 + 7.1 + 9.7$ ______ V

17. $88.7 + 0.6$ ______ F

18. $26.2 + 9.56$ ______ I

19. $2.502 + 93.87$ ______ K

20. $97 + 12.81$ ______ C

___ R ___ ___ ___ ___ ___ S
76.8 131.8 109.81 96.372 80.00 12.66

___ ___ ___ ___ ___ L ___ ___
60.26 68.233 21.4 65.1 32.5 8.017 3.7

___ ___ ___ ___ ___ ___ ___ !
13.443 89.3 103.11 0.9 35.76 25.58 1.3

NAME

SHARPEN YOUR SKILLS

Subtracting Decimals

What state is called "the state where the sun spends the winter"?

Subtract. Find each answer in the table. Cross out the letter next to it. The remaining letters spell the answer.

W	10.95
C	3.16
Y	42.763
A	356.03
~~X~~	0.22
I	4.798
R	18.792
O	2.16
M	6.269
F	9.51
O	527.16
V	68.91
I	2.165
N	0.564
E	0.139
R	9.44
M	0.17
A	4.32
Z	83.6
O	1.002
N	23.81
M	8.289
N	50.17
O	512.35
A	0.82
U	62.507
K	1.811
N	0.191
S	0.424
A	450.73
T	2.31

1. 0.92 − 0.7 ______

2. 9.81 − 0.30 ______

3. 5.40 − 3.24 ______

4. 6.65 − 2.33 ______

5. 7.62 − 6.80 ______

6. 28.25 − 17.30 ______

7. 81.31 − 57.5 ______

8. 572.90 − 45.74 ______

9. 620.6 − 108.25 ______

10. 0.624 − 0.200 ______

11. 6.849 − 0.58 ______

12. 99.511 − 37.004 ______

13. 57.39 − 14.627 ______

14. 23.719 − 15.43 ______

15. 4.501 − 2.69 ______

16. 5.398 − 0.6 ______

17. 19.82 − 19.681 ______

18. 6.281 − 6.09 ______

19. 0.67 − 0.5 ______

20. 5.21 − 2.9 ______

21. 86.51 − 17.6 ______

22. 0.724 − 0.16 ______

23. 9.74 − 0.3 ______

24. 40.8 − 37.64 ______

___ ___ ___ ___ ___ ___ ___

SHARPEN YOUR SKILLS

Extending Place Value

For each number, name the value of the last decimal place.

1. 356.890 ______
2. 21.0954 ______
3. 567.82435 ______
4. 27.09865 ______
5. 376.5416 ______
6. 46.9003 ______

For each number, name the place that contains a 7.

7. 8.4507 ______
8. 45.7692 ______
9. 3.8971 ______
10. 34.8791 ______
11. 713.98 ______
12. 76.0005 ______
13. 257.52 ______
14. 2.45897 ______
15. 41.1744 ______

Solve each problem.

16. What would your calculator display show if you entered six hundred and fifty-six ten-thousandths? ______

17. What would your calculator display show if you entered thirty-four and four thousand three hundred twenty-seven hundred-thousandths? ______

NAME

SHARPEN YOUR SKILLS

Multiplying a Decimal by a Whole Number

Multiply to complete the cross-number puzzle. The decimal points have been placed in the puzzle.

Across

1. 2×3.16

3. 52×3.4

5. 24×0.18

7. 6×0.9542

8. 5×8.103

10. 4×9.18

13. 2×2.7036

Down

1. 7×0.9312

2. 33×0.735

4. 8×0.8073

6. 15×1.71

9. 45×18.5

11. 24×30.6

12. 7×9.06

SHARPEN YOUR SKILLS

Choose an Operation

Tell which operation to use. Then solve each problem.

1. Mr. Collins bought 6 apple trees. How much money did he spend?

____

2. The Walkers bought 3 dozen tulip bulbs. How much money did they spend?

____

3. Mrs. Rodríguez bought 1 bag of fertilizer and 1 box of grass seed. How much money did she spend?

____

4. Mr. Cressman planted 240 marigolds in his yard. A frost destroyed 49 of the plants. How many marigolds survived the frost?

____

5. Mr. Lee bought 32 geranium plants for $0.49 each. How much money did he spend?

____

6. Ms. Mitchell spent $10.80. How much change did she receive from $20?

____

7. Three families shared the cost of a box of ground cover equally. What was the cost per family?

____

8. James and Josie bought a dozen tulip bulbs. They shared the cost equally. How much did each of them spend?

____

Use after pages 256–257.

NAME

SHARPEN YOUR SKILLS

Mental Math for Multiplying by 10, 100, 1,000

Find each product. Use mental math.

1. $10 \times 2.48 =$ ______________

2. $100 \times 2.48 =$ ______________

3. $11.236 \times 100 =$ ______________

4. $11.266 \times 10 =$ ______________

5. $0.6 \times 10 =$ ______________

6. $0.6 \times 1{,}000 =$ ______________

7. $2.5 \times 100 =$ ______________

8. $0.319 \times 1{,}000 =$ ______________

Multiply across. Then multiply down.

9. ×

0.03	10	
100	4.24	

10.

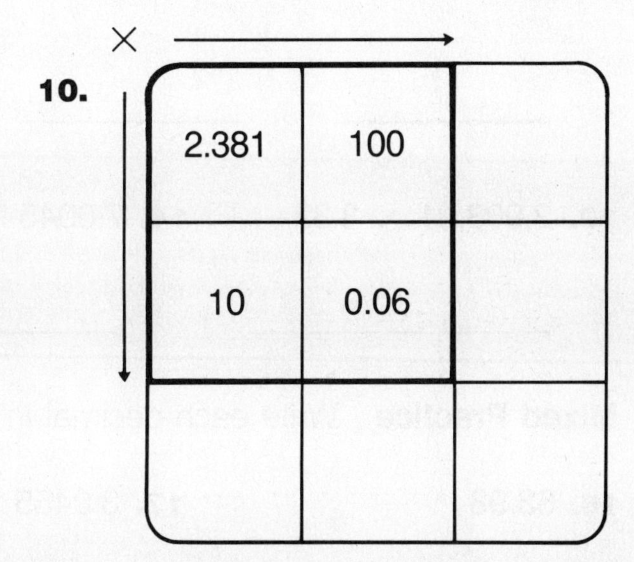

11.

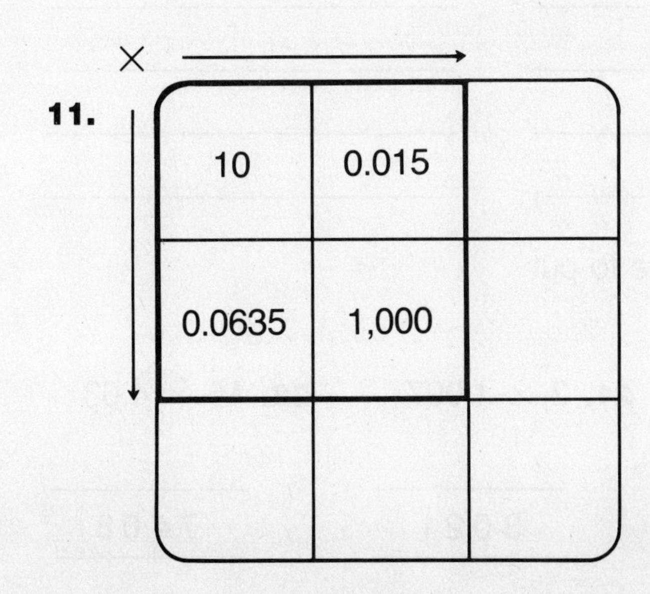

12.

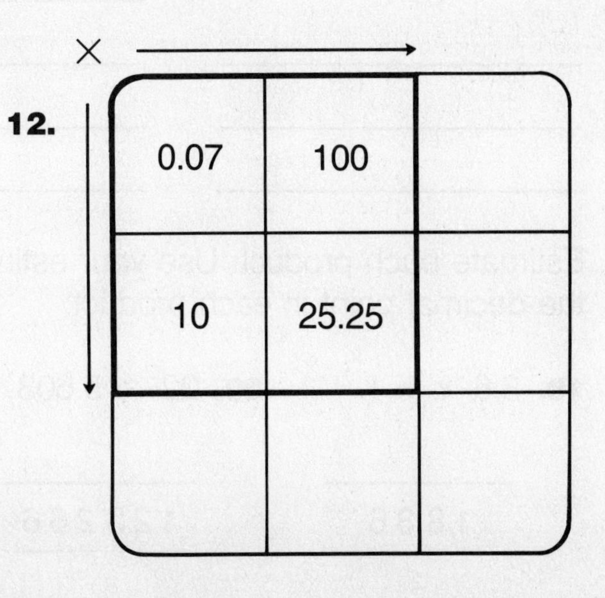

SHARPEN YOUR SKILLS

Estimating Products

Estimate the products.

1. 5.99 × 3.1 ______

2. 12.2 × 4.8 ______

3. 6.75 × 3.12 ______

4. 3.42 × 49.89 ______

5. 44.03 × 2.22 ______

6. 24.7 × 1.2 ______

7. 3.07 × 5.6 ______

8. 0.88 × 5.989 ______

9. 39.4 × 12.22 ______

10. 224.8 × 10.1 ______

11. 33.03 × 5.94 ______

12. 3.34 × $20 ______

13. 2,003.21 × 3.32 ______

14. 7.0045 × $0.25 ______

15. 45.67 × 9.97 ______

Mixed Practice Write each decimal in words.

16. 63.98 ______

17. 3.0455 ______

18. 12.32560 ______

Estimate each product. Use your estimate to put the decimal point in each product.

19. 3.6 × 5.1 ______ 1836

20. 22 × 5.603 ______ 123266

21. 3 × 1.007 ______ 3021

22. 16 × 4.63 ______ 7408

NAME

SHARPEN YOUR SKILLS

Multiplying by a Decimal

Multiply.

1. 0.3×0.6

2. 5×0.5

3. 4×0.06

4. 27×0.8

5. 0.91×0.9

6. 0.7×0.7

7. 6.3×0.04

8. 81×0.06

9. 4.71×0.3

10. 1.63×0.07

11. 0.82×0.4

12. 14.5×0.05

13. $5.628 \times 0.7 =$ ______________

14. $31.27 \times 0.03 =$ ______________

15. $12.4 \times 7 =$ ______________

16. $1.6 \times 2.4 =$ ______________

Solve each problem.

17. A Japanese maple tree can grow to be 6 meters tall. A paperbark maple tree is about 1.3 times as tall. About how tall is a paperbark maple tree?

____

18. A queen palm tree can grow to be 15 meters tall. A silk tree can grow about 0.8 times as tall. How tall can a silk tree grow?

____

Use after pages 264–265.

NAME

SHARPEN YOUR SKILLS

Multiplying Decimals: Zeros in the Product

Why is the letter "A" like clover?

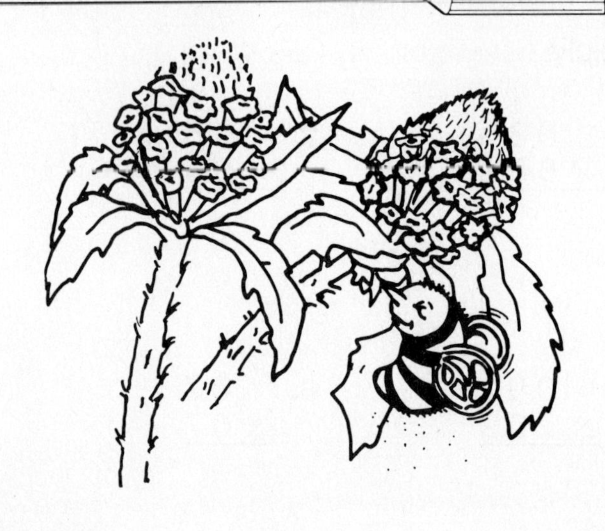

To solve the riddle, multiply. Then draw a line from each exercise to its answer. Write the letter in the box that has the same number as the exercise.

1. 0.3 × 0.07	• **E** 0.048
2. 2.2 × 0.004	• **M** 0.0216
3. 0.03 × 0.9	• **A** 0.01735
4. 6 × 0.008	• **E** 0.027
5. 0.452 × 0.03	• **E** 0.0035
6. 0.76 × 0.1	• **T** 0.0712
7. 3.6 × 0.006	• **A** 0.021
8. 0.005 × 0.7	• **O** 0.076
9. 2.8 × 0.03	• **R** 0.05226
10. 3.47 × 0.005	• **S** 0.084
11. 1.359 × 0.008	• **B** 0.0088
12. 3.56 × 0.02	• **F** 0.010872
13. 2.07 × 0.04	• **C** 0.01356
14. 2.613 × 0.02	• **E** 0.0828

Because

2.	3.	4.

5.	6.	7.	8.	9.

10.	11.	12.	13.	14.

it.

NAME

SHARPEN YOUR SKILLS

Exploring Division of Decimals

Divide. **Remember** to put the decimal point in the quotient.

1. 2.52 ÷ 3 ______

2. 15.6 ÷ 6 ______

3. 1.407 ÷ 7 ______

4. 430.5 ÷ 5 ______

5. 70.29 ÷ 9 ______

6. 2.052 ÷ 6 ______

7. 410.2 ÷ 7 ______

8. 5.584 ÷ 8 ______

9. 0.275 ÷ 5 ______

10. 252.4 ÷ 4 ______

11. 67.68 ÷ 9 ______

12. 0.772 ÷ 8 ______

Solve each problem. You may use play money to do these exercises.

13. There were 6 students working in the book booth at the school fair. Together they made $47.34. Each was given an equal share. How much did each student make?

______

14. Linda worked on Thursday for 3 hours and Friday for 4 hours. Wendy worked on Friday for 7 hours. They earned the same amount per hour and were paid $77.14 total. How much money did each one make?

______

NAME

Multiple Step Problems

CUCUMBERS 29¢ each
CARROTS 39¢ per bag
TOMATOES 69¢ per pound
PEPPERS 30¢ each
ONIONS 39¢ per pound
LETTUCE 59¢ per head
BROCCOLI 79¢ per pound
POTATOES $2.69 per bag

Solve each problem.

1. Conchita bought 1 pound of tomatoes, 1 head of lettuce, and 5 pounds of onions. How much did she spend?

____

2. Harold bought 1 bag of potatoes, 2 pounds of broccoli, and 1 pepper. What was the total cost?

____

3. Andrea bought 3 bags of carrots and 4 cucumbers. How much did she pay?

____

4. Scott bought 2 pounds of tomatoes. Amy bought a bag of potatoes. How much more did Amy spend than Scott?

____

For each purchase, find the change from $10.

5. 6 peppers

____

6. 4 pounds of tomatoes

____

7. 1 head of lettuce and 1 bag of potatoes

____

8. 1 bag of carrots and 1 pound of broccoli

____

NAME

SHARPEN YOUR SKILLS

Patterns

Draw the next three figures in each pattern.

1.

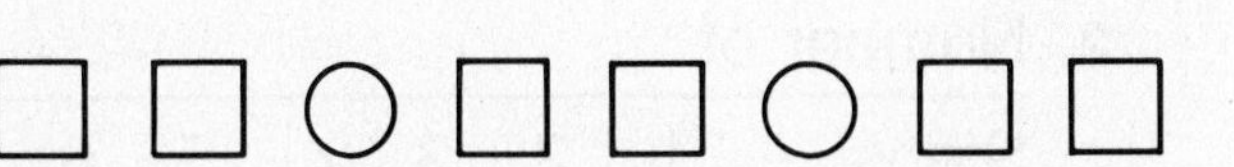

2.

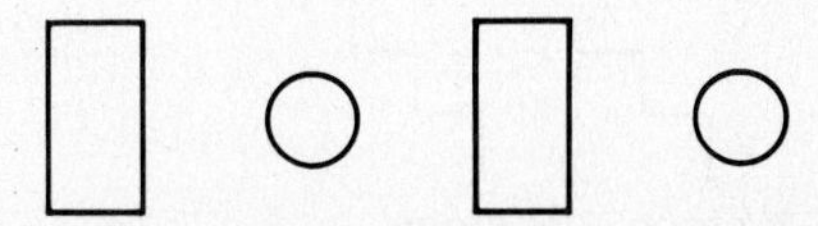

3.

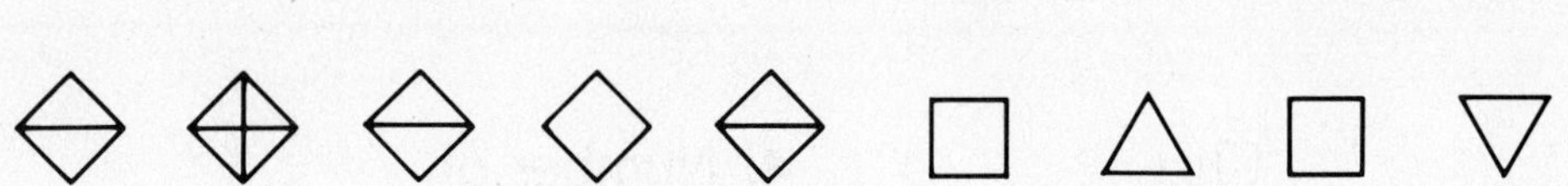

4.

5.

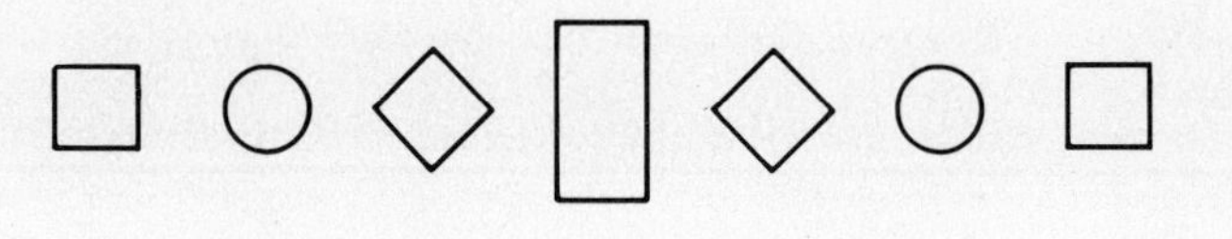

6.

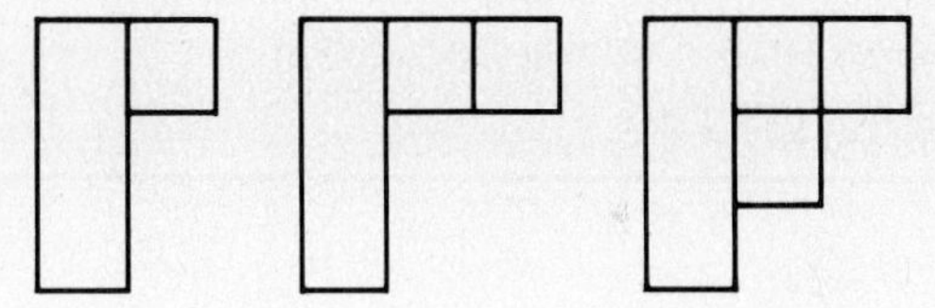

7. Use the following figures to create your own pattern. Ask a classmate to try to find the pattern. You may use any of the figures as many times as you like. You do not have to use them all.

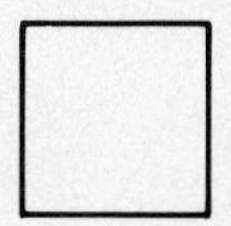 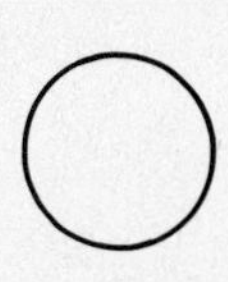 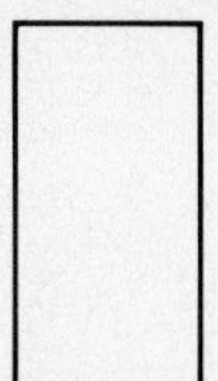 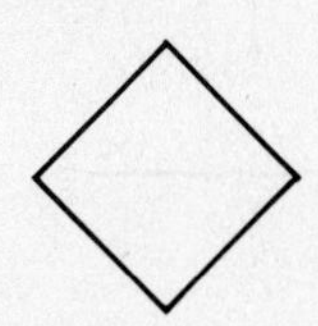 

SHARPEN YOUR SKILLS

Number Patterns

Draw the next three figures in each pattern.
Complete each table.

1.

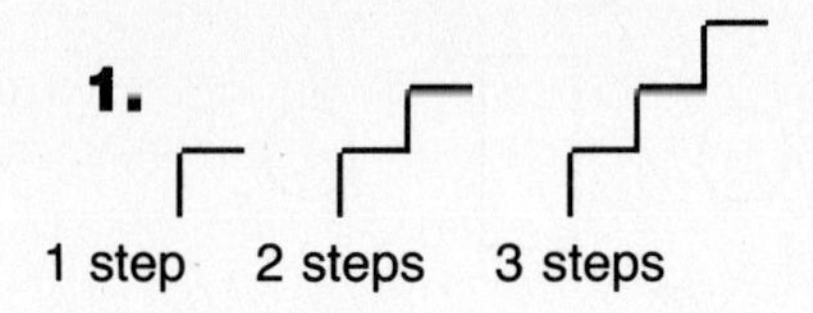

2. Number of

Steps	1	2	3	4	5	6
Lines	2	4	6			

3.

1 row 2 rows 3 rows

4. Number of

Rows	1	2	3	4	5	6
Squares	2	6				

5.

1 line 2 lines 3 lines

6. Number of

Lines	1	2	3	4	5	6
Triangles that do not overlap	2	4				

Write the next three numbers in each sequence.

7. 1, 2, 4, 8, ______, ______, ______

8. 1, 3, 5, ______, ______, ______

9. 4, 8, 12, ______, ______, ______

NAME

SHARPEN YOUR SKILLS

Number Relationships

Describe the pattern in each sequence.
Then write the next four numbers.

1. 3, 5, 7, 9, 11, . . .

2. 85, 77, 69, 61, 53, . . .

3. 2, 9, 18, 25, 50, . . .

4. 2, 10, 5, 25, 20, . . .

5. 18, 17, 21, 20, 24, 23, . . .

6. 50, 40, 400, 390, 3,900, . . .

7. 98, 90, 93, 85, 88, . . .

8. 1, 1.5, 2.0, 2.5, 3.0, . . .

Critical Thinking Look at the 5-by-5 array.

1	2	3	4	5
6	7	8	9	10
11	12	13	14	15
16	17	18	19	20
21	22	23	24	25

9. Write the sequence of numbers on the diagonal from upper left to lower right. Describe the pattern.

10. Write the sequence of numbers on the diagonal from lower left to upper right. Describe the pattern.

NAME

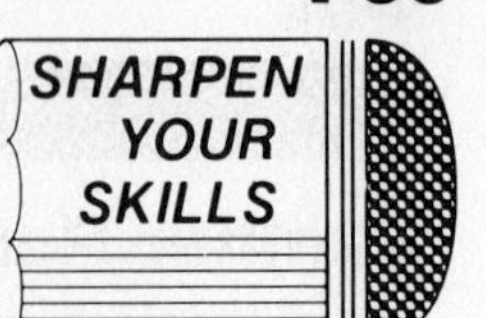

Find a Pattern

There are 8 people in a room. Each one holds a conversation with each other person in the room one time. Complete the table to find how many conversations take place.

1.

Number of People	Number of Conversations
2	1
3	3
4	______
5	______
6	______
7	______
8	______

2. How many conversations would take place if there were 12 people in the room?

____

In a checkers tournament, each player plays one other player in the first round. The winners play each other in the second round. Those winners play each other in the third round, and so on, until there is a winner.

3. How many rounds would there be if there were 8 players in the beginning?

____

4. How many rounds would there be if there were 16 players in the beginning?

____

5. How many games would there be if there were 16 players at the beginning?

____

6. How many games would there be if there were 32 players at the beginning?

____

Critical Thinking Look at the answers to Exercises 5 and 6. What pattern do you see?

____

____

____

NAME

SHARPEN YOUR SKILLS

Locating Points with Ordered Pairs

Use the grid below. Plot the points located by the ordered pairs in Exercise 1. Connect the points in the order that you make them. Repeat the process for each exercise to find the answer to the question.

1. (24, 1) (24, 7) (27, 7) (27, 4) (26, 4) (24, 4) (27, 1)

2. (20, 7) (20, 1) (23, 1) (23, 7)

3. (16, 7) (16, 1) (19, 1) (19, 7) (16, 7)

4. (11, 7) (8, 7) (8, 4) (11, 4) (11, 1) (8, 1)

5. (7, 7) (7, 1)

6. (15, 7) (12, 7) (12, 4) (15, 4) (15, 1) (12, 1)

7. (28, 7) (28, 1)

8. (2, 1) (2, 7) (4, 4) (6, 7) (6, 1)

What is the name of the "Show Me" state?

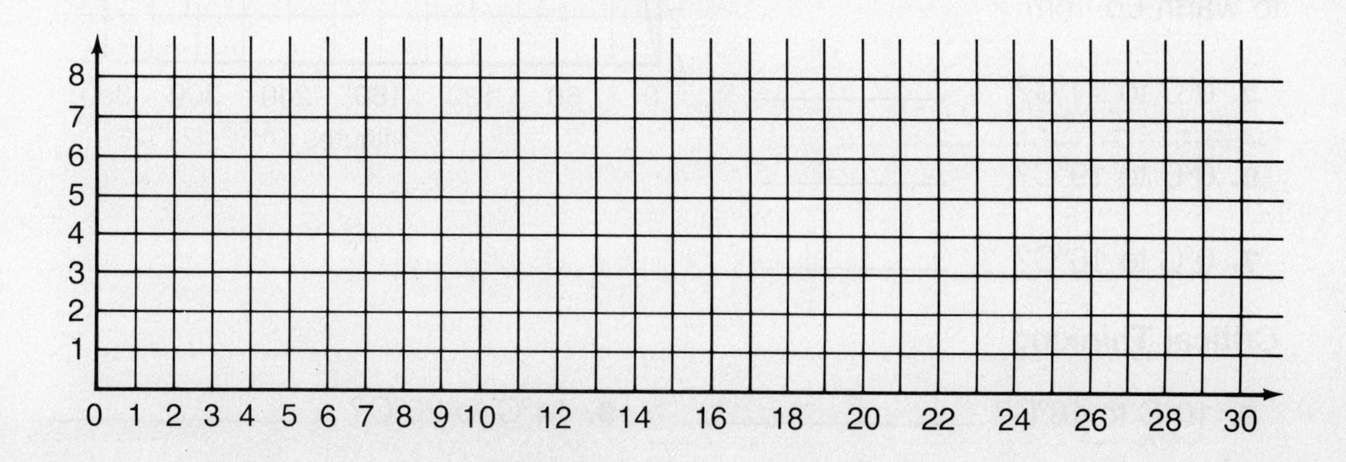

Use after pages 292–293.

NAME

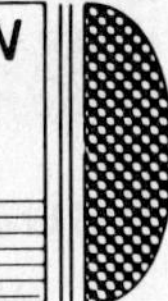

Use Data from a Graph

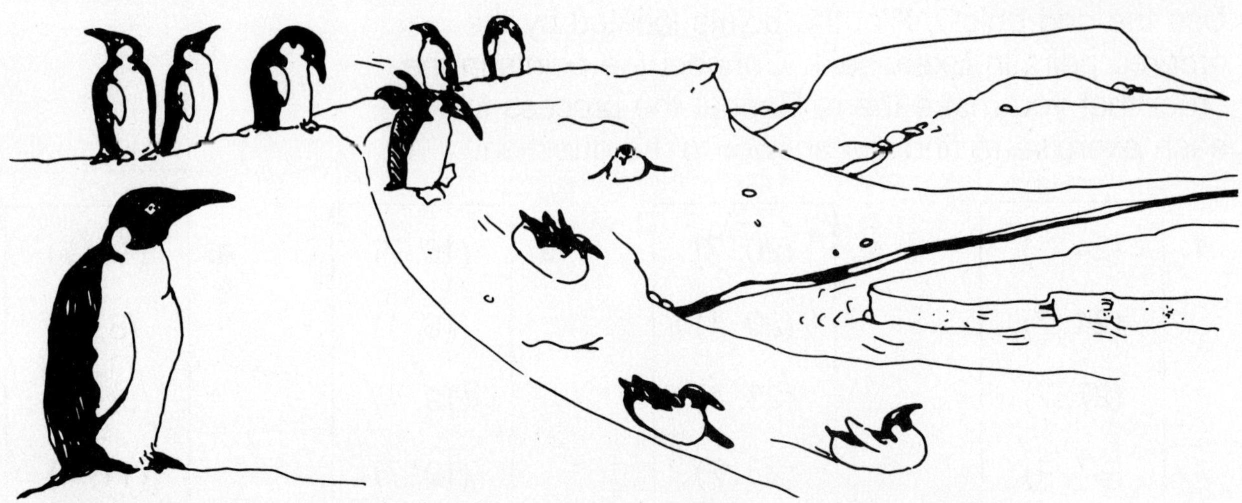

The graph shows the time it takes for the water in an ice cube to melt and warm up to room temperature (20°C).

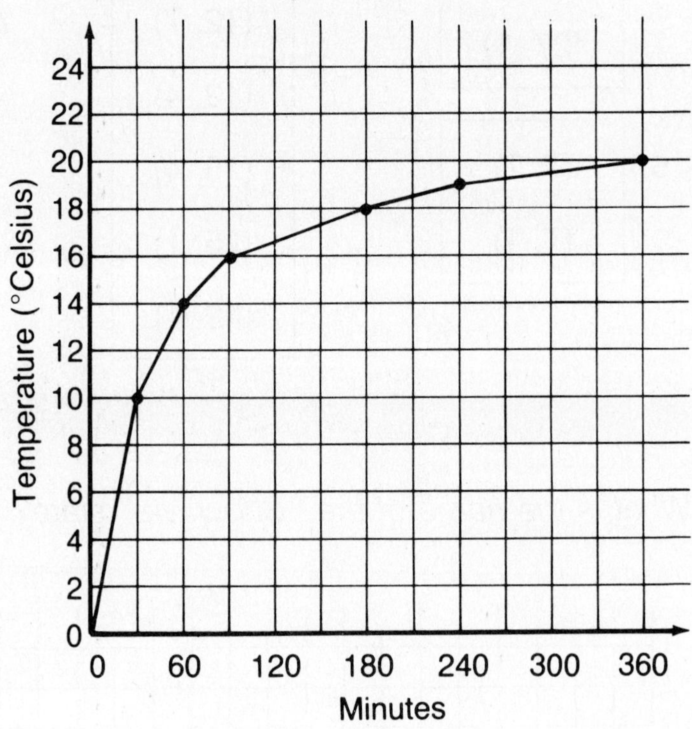

What is the temperature of the water at

1. 0 minutes? ________

2. 360 minutes? ________

3. 150 minutes? ________

4. 30 minutes? ________

How many minutes does it take for the water to warm up from

5. 0°C to 14°C? ________

6. 0°C to 19°C? ________

7. 0°C to 10°C? ________

Critical Thinking

8. 10°C to 18°C? ____________

9. 14°C to 20°C? ____________

NAME

SHARPEN YOUR SKILLS

Number Relationships and Ordered Pairs

Find the missing numbers.

1. Rule: $A \times 10 = B$

A	B
3	___
4	___
5	___
___	60

2. Rule: $A + 12 = B$

A	B
7	___
9	___
11	___
___	25

3. Rule: $A - 9 = B$

A	B
28	___
26	___
___	15
22	___

Find the missing numbers. Write the rule.

4. Rule: ____________

A	B
1	15
2	30
3	___
___	60

5. Rule: ____________

A	B
100	___
90	83
80	73
70	___

6. Rule: ____________

A	B
10	60
20	___
30	___
40	90

Make the table to show the relationship between time and distance for the following items. Include a column for the ordered pairs.

7. A person walking at 3 miles per hour for 5 hours

8. A horse running at 10 miles per hour for 5 hours

9. On a separate piece of graph paper, graph the ordered pairs in Exercises 7 and 8.

SHARPEN YOUR SKILLS

Fractions

Write the fraction to name the part that is shaded.

1. 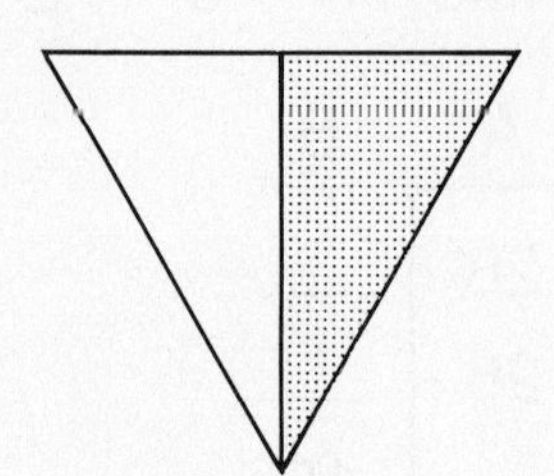 ______

2. 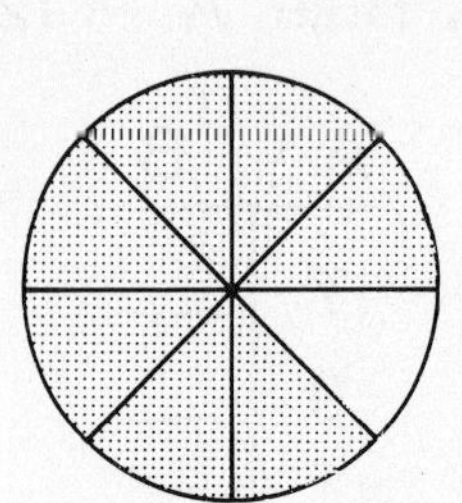 ______

3. 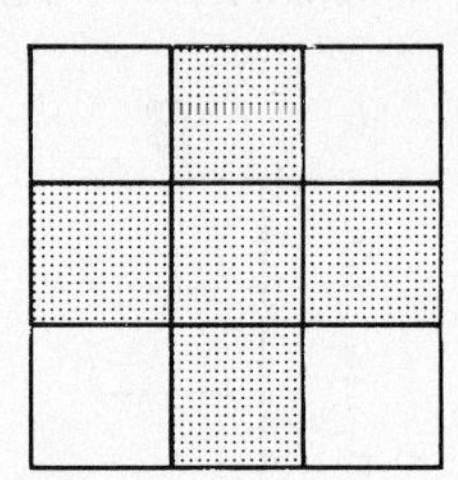 ______

Choose the figure that illustrates the fraction.

4. Two thirds

5. Five sixths

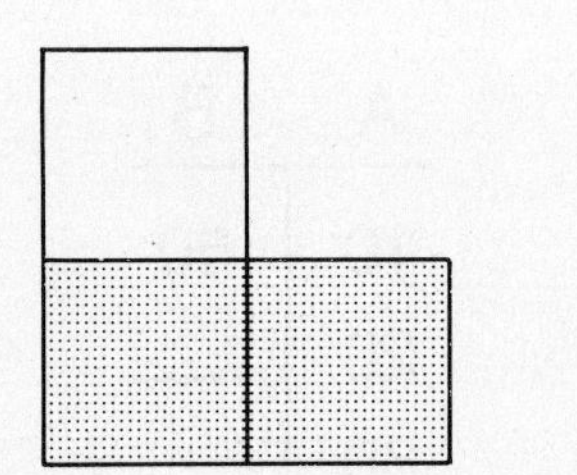

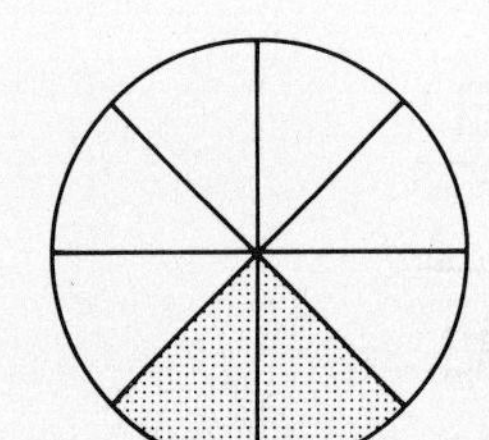

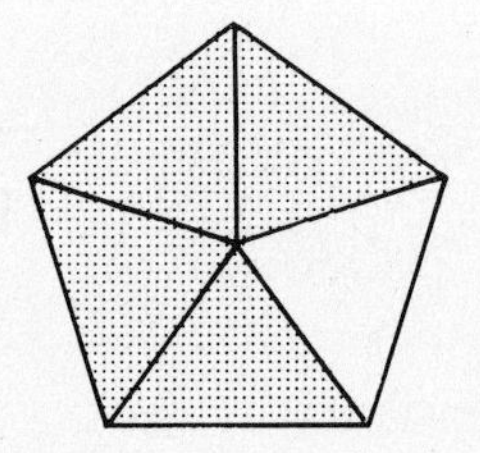

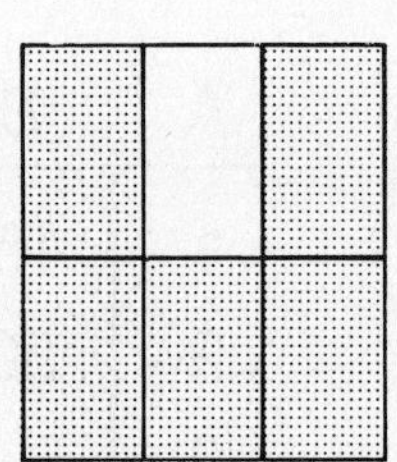

Write the fraction that names the points labeled *A*.

6. 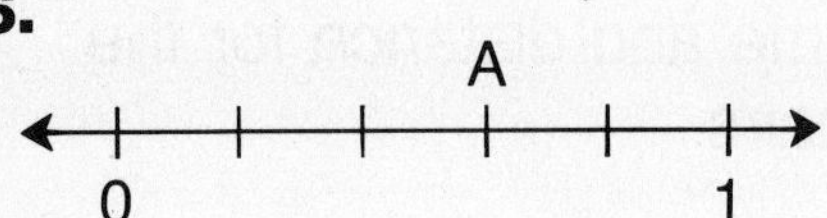

______

7. 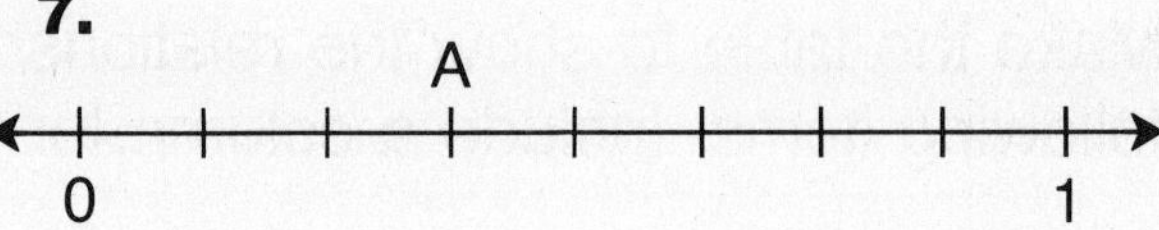

______

Write each fraction using numerals.

8. Three fifths ______

9. One fourth ______

10. Three thirds ______

11. Two sixths ______

12. Seven twelfths ______

13. Four ninths ______

14. Two sevenths ______

15. Zero eighths ______

NAME

SHARPEN YOUR SKILLS

Equal Fractions

Multiply to find equal fractions.

1. $\frac{1}{4} = \frac{\square}{12}$

2. $\frac{2}{5} = \frac{\square}{15}$

3. $\frac{3}{7} = \frac{\square}{21}$

4. $\frac{1}{8} = \frac{\square}{48}$

5. $\frac{5}{12} = \frac{\square}{24}$

6. $\frac{5}{6} = \frac{\square}{12}$

Divide to find equal fractions.

7. $\frac{6}{12} = \frac{\square}{2}$

8. $\frac{7}{21} = \frac{\square}{3}$

9. $\frac{8}{40} = \frac{\square}{5}$

10. $\frac{7}{42} = \frac{1}{\square}$

11. $\frac{12}{24} = \frac{1}{\square}$

12. $\frac{10}{45} = \frac{\square}{9}$

Mixed Practice Multiply or divide to find equal fractions.

13. $\frac{1}{5} = \frac{\square}{10}$

14. $\frac{4}{10} = \frac{\square}{5}$

15. $\frac{2}{4} = \frac{1}{\square}$

16. $\frac{3}{5} = \frac{\square}{10}$

17. $\frac{4}{8} = \frac{\square}{4}$

18. $\frac{3}{9} = \frac{\square}{3}$

19. $\frac{1}{4} = \frac{\square}{8}$

20. $\frac{10}{10} = \frac{5}{\square}$

21. $\frac{8}{10} = \frac{\square}{5}$

22. $\frac{3}{4} = \frac{\square}{8}$

23. $\frac{2}{8} = \frac{1}{\square}$

24. $\frac{5}{5} = \frac{\square}{10}$

Tell whether the fractions are equal.

25. $\frac{2}{9}$ $\frac{3}{18}$ ________

26. $\frac{3}{7}$ $\frac{18}{42}$ ________

27. $\frac{5}{6}$ $\frac{6}{30}$ ________

28. $\frac{9}{10}$ $\frac{90}{90}$ ________

29. $\frac{1}{5}$ $\frac{9}{45}$ ________

30. $\frac{4}{8}$ $\frac{6}{12}$ ________

Use after pages 310–311.

SHARPEN YOUR SKILLS

Fractions in Lowest Terms

Cross out each box that contains a fraction in lowest terms.
The remaining letters contain a hidden message.

$\frac{3}{6}$ F	$\frac{3}{4}$ P	$\frac{3}{9}$ R	$\frac{1}{3}$ A	$\frac{5}{10}$ A	$\frac{7}{8}$ E	$\frac{6}{18}$ C	$\frac{2}{10}$ T	$\frac{4}{12}$ I	$\frac{6}{8}$ O	$\frac{9}{18}$ N	$\frac{3}{6}$ S

$\frac{4}{15}$ C	$\frac{9}{13}$ R	$\frac{6}{7}$ O	$\frac{7}{21}$ A	$\frac{2}{5}$ S	$\frac{1}{11}$ N	$\frac{10}{25}$ R	$\frac{3}{7}$ O	$\frac{5}{12}$ B	$\frac{2}{16}$ E	$\frac{6}{23}$ G	$\frac{1}{8}$ Y

$\frac{1}{7}$ I	$\frac{2}{5}$ X	$\frac{9}{81}$ F	$\frac{5}{8}$ F	$\frac{1}{9}$ V	$\frac{5}{14}$ T	$\frac{2}{24}$ U	$\frac{3}{4}$ R	$\frac{12}{42}$ N	$\frac{1}{13}$ P	$\frac{3}{10}$ B	$\frac{8}{15}$ H

.

Write each fraction in lowest terms.

1. $\frac{16}{20}$ = ______ **2.** $\frac{3}{27}$ = ______ **3.** $\frac{24}{64}$ = ______ **4.** $\frac{5}{20}$ = ______

5. $\frac{21}{30}$ = ______ **6.** $\frac{27}{45}$ = ______ **7.** $\frac{6}{9}$ = ______ **8.** $\frac{10}{12}$ = ______

9. $\frac{10}{15}$ = ______ **10.** $\frac{4}{18}$ = ______ **11.** $\frac{8}{14}$ = ______ **12.** $\frac{15}{20}$ = ______

13. $\frac{12}{30}$ = ______ **14.** $\frac{14}{21}$ = ______ **15.** $\frac{36}{45}$ = ______ **16.** $\frac{9}{15}$ = ______

17. $\frac{24}{32}$ = ______ **18.** $\frac{9}{36}$ = ______ **19.** $\frac{16}{24}$ = ______ **20.** $\frac{30}{36}$ = ______

21. $\frac{35}{42}$ = ______ **22.** $\frac{18}{27}$ = ______ **23.** $\frac{10}{16}$ = ______ **24.** $\frac{40}{64}$ = ______

25. $\frac{15}{27}$ = ______ **26.** $\frac{30}{35}$ = ______ **27.** $\frac{21}{28}$ = ______ **28.** $\frac{18}{24}$ = ______

SHARPEN YOUR SKILLS

Mixed Numbers

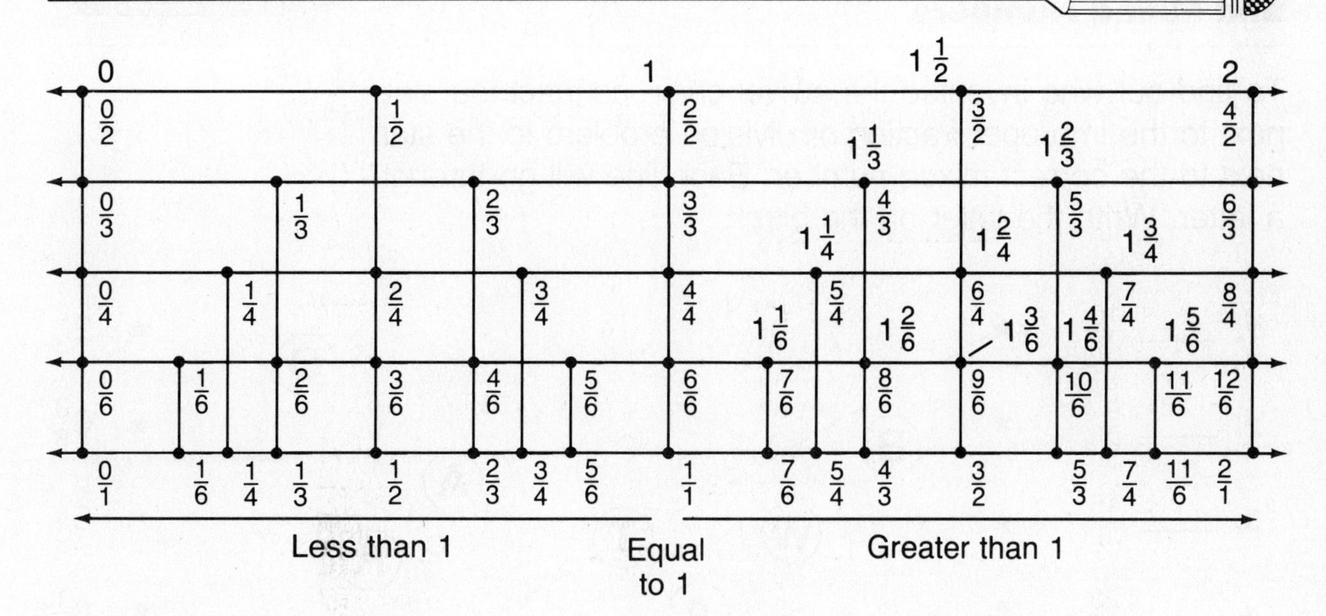

Use the number lines to answer the following questions. **Remember** that one point on a number line can have more than one name.

1. What fraction is equal to $\frac{2}{6}$? ________

2. What mixed number is equal to $\frac{9}{6}$? ________

3. Which fraction is less, $\frac{7}{6}$ or $\frac{5}{4}$? ________

4. What whole number is equal to $\frac{8}{4}$? ________

5. What mixed number in fourths is closest to $1\frac{1}{6}$? ________

6. Which mixed number is more, $1\frac{3}{4}$ or $1\frac{4}{6}$? ________

7. Amy wants to cut sections that are $\frac{1}{4}$ in. long from one length of ribbon. If she can cut 7 sections from the length, how long is the original ribbon?

____

8. Jordan cut a wooden dowel rod into 11 equal pieces. If each piece is $\frac{1}{6}$ inch long, how long was the original rod?

____

NAME

SHARPEN YOUR SKILLS

Improper Fractions, Quotients, and Mixed Numbers

To find out who invented the swivel chair, connect the star next to the improper fraction or division problem to the star next to the correct mixed number. Each line will go through a letter. Write the letter on the blank.

1. ____	$\frac{10}{3}$	★	★	$2\frac{2}{3}$
2. ____	$\frac{9}{2}$	★	★	$6\frac{3}{8}$
3. ____	$\frac{16}{6}$	★	★	$6\frac{1}{3}$
4. ____	$\frac{23}{4}$	★	★	$8\frac{1}{4}$
5. ____	$8\overline{)51}$	★	★	$10\frac{1}{3}$
6. ____	$\frac{68}{7}$	★	★	$3\frac{1}{3}$
7. ____	$6\overline{)38}$	★	★	$4\frac{1}{7}$
8. ____	$\frac{29}{7}$	★	★	$6\frac{1}{12}$
9. ____	$\frac{66}{8}$	★	★	$9\frac{7}{9}$
10. ____	$10\overline{)35}$	★	★	$4\frac{1}{2}$
11. ____	$\frac{31}{3}$	★	★	$3\frac{1}{2}$
12. ____	$9\overline{)88}$	★	★	$10\frac{1}{10}$
13. ____	$\frac{76}{25}$	★	★	$5\frac{3}{4}$
14. ____	$\frac{101}{10}$	★	★	$3\frac{1}{25}$
15. ____	$12\overline{)73}$	★	★	$9\frac{5}{7}$

O B A W T R A J V H K F M E E P L M S N F W D R N O S G

NAME

SHARPEN YOUR SKILLS

Fractions and Estimation

For each fraction, tell if the fraction is less than $\frac{1}{2}$, equal to $\frac{1}{2}$, or greater than $\frac{1}{2}$.

1. $\frac{7}{9} =$ ______ **2.** $\frac{1}{6} =$ ______ **3.** $\frac{5}{10} =$ ______ **4.** $\frac{1}{10} =$ ______

5. $\frac{12}{24} =$ ______ **6.** $\frac{4}{25} =$ ______ **7.** $\frac{17}{20} =$ ______ **8.** $\frac{14}{28} =$ ______

9. $\frac{6}{100} =$ ______ **10.** $\frac{50}{100} =$ ______ **11.** $\frac{13}{50} =$ ______ **12.** $\frac{13}{25} =$ ______

13. $\frac{16}{32} =$ ______ **14.** $\frac{99}{100} =$ ______ **15.** $\frac{17}{32} =$ ______ **16.** $\frac{13}{26} =$ ______

Estimate whether each sum is (a) less than 1, (b) greater than 1, or (c) greater than the larger fraction.

17. $\frac{1}{3} + \frac{1}{4}$ ______ **18.** $\frac{2}{3} + \frac{3}{4}$ ______ **19.** $\frac{1}{6} + \frac{4}{5}$ ______

20. $\frac{1}{12} + \frac{1}{10}$ ______ **21.** $\frac{8}{9} + \frac{11}{12}$ ______ **22.** $\frac{3}{20} + \frac{7}{12}$ ______

23. $\frac{1}{16} + \frac{24}{25}$ ______ **24.** $\frac{15}{16} + \frac{19}{20}$ ______ **25.** $\frac{3}{20} + \frac{1}{18}$ ______

26. $\frac{7}{25} + \frac{1}{50}$ ______ **27.** $\frac{3}{100} + \frac{31}{50}$ ______ **28.** $\frac{79}{100} + \frac{43}{50}$ ______

NAME

SHARPEN YOUR SKILLS

Work Backward

Solve each problem.

1. Robert is saving aluminum cans for the school recycling project. He saved twice as many as Lisa, three times as many as Jean, and half as many as Troy. Troy had 48. How many did each student collect?

Troy: _____ Robert: _____

Lisa: _____ Jean: _____

2. Barry collected the same number of cans as Barbara. Janet collected six times as many as Barry, and Juan got 100, which was ten times more than Barry and Barbara each got. How many did each student collect?

Barry: _____ Barbara: _____

Juan: _____ Janet: _____

3. For three weeks the football team practiced each school day. Robert attended every practice. John attended all but one practice, and Sid attended only half of the practices that John did. How many sessions did each player attend?

Robert: _____ John: _____

Sid: _____

4. The football season had 10 games. Five games were played at home, and 5 games were played at other schools. Of the games played at home, 1 was lost and 1 was tied. Of the games played away, 2 were lost and 1 was tied. How many games were lost, tied, and won?

Lost: _____ Tied: _____

Won: _____

5. The track team ran in 12 events each week. Each week the team won 3 more events than in the previous week. They won 7 events in the third week. How many events did they win in the three weeks?

____

6. The baseball team lost $\frac{1}{2}$ as many games in the second week as in the first week. They lost $\frac{1}{3}$ as many games in the third week as in the second week. If they lost 1 game the third week, how many games did they lose in the three weeks?

____

NAME

SHARPEN YOUR SKILLS

Finding the Least Common Demoninator

Write the fractions with the least common denominator.

1. $\frac{3}{8}, \frac{1}{4}$ ______
2. $\frac{2}{3}, \frac{5}{6}$ ______
3. $\frac{3}{10}, \frac{4}{5}$ ______
4. $\frac{3}{5}, \frac{1}{2}$ ______
5. $\frac{3}{4}, \frac{1}{3}$ ______
6. $\frac{2}{5}, \frac{2}{3}$ ______
7. $\frac{2}{5}, \frac{7}{10}$ ______
8. $\frac{7}{12}, \frac{3}{4}$ ______
9. $\frac{2}{3}, \frac{5}{24}$ ______
10. $\frac{1}{2}, \frac{3}{7}$ ______
11. $\frac{3}{4}, \frac{1}{5}$ ______
12. $\frac{1}{3}, \frac{7}{8}$ ______
13. $\frac{5}{12}, \frac{1}{3}, \frac{1}{4}$ ______
14. $\frac{1}{2}, \frac{5}{6}, \frac{2}{3}$ ______
15. $\frac{3}{5}, \frac{1}{4}, \frac{2}{5}$ ______
16. $\frac{2}{3}, \frac{11}{12}, \frac{3}{4}$ ______
17. $\frac{3}{10}, \frac{4}{5}, \frac{1}{2}$ ______
18. $\frac{5}{6}, \frac{1}{18}, \frac{1}{3}$ ______

Critical Thinking How many common denominators do you think any two fractions can have? Why?

______

______

NAME

SHARPEN YOUR SKILLS

Comparing and Ordering Fractions

Compare the fractions, whole numbers, and mixed numbers. Use <, >, or =.
Remember to find the least common denominator.

1. $\frac{5}{6} \bigcirc \frac{6}{6}$

2. $\frac{3}{4} \bigcirc \frac{2}{4}$

3. $2\frac{2}{5} \bigcirc 1\frac{2}{5}$

4. $\frac{5}{6} \bigcirc \frac{1}{8}$

5. $\frac{3}{8} \bigcirc \frac{9}{24}$

6. $5\frac{9}{10} \bigcirc 5\frac{7}{20}$

7. $6\frac{11}{12} \bigcirc 6\frac{1}{6}$

8. $1\frac{7}{15} \bigcirc 1\frac{3}{5}$

9. $13\frac{2}{9} \bigcirc 13\frac{5}{9}$

10. $25\frac{3}{5} \bigcirc 25\frac{6}{10}$

11. $33\frac{4}{7} \bigcirc 33\frac{2}{3}$

12. $56\frac{1}{4} \bigcirc 56\frac{3}{8}$

13. $\frac{12}{12} \bigcirc \frac{0}{12}$

14. $\frac{16}{32} \bigcirc \frac{20}{40}$

15. $\frac{21}{20} \bigcirc \frac{19}{20}$

16. $\frac{50}{100} \bigcirc \frac{5}{100}$

17. $\frac{18}{6} \bigcirc \frac{81}{6}$

18. $\frac{25}{5} \bigcirc \frac{52}{5}$

19. $\frac{7}{9} \bigcirc \frac{4}{3}$

20. $22\frac{11}{24} \bigcirc 22\frac{1}{2}$

21. $16\frac{3}{6} \bigcirc 16\frac{6}{12}$

Write the fractions, whole numbers, and mixed numbers in order from least to greatest.

22. $\frac{4}{15}, \frac{1}{5}, \frac{7}{30}$ ____________________

23. $3, 2\frac{11}{16}, 2\frac{1}{2}$ ____________________

24. $1\frac{15}{18}, 1\frac{7}{9}, 1\frac{2}{3}$ ____________________

25. $2\frac{3}{4}, 2\frac{11}{12}, 2\frac{5}{6}$ ____________________

Use after pages 326–327.

NAME

SHARPEN YOUR SKILLS

Fractions and Decimals

Write each decimal as a fraction or a mixed number.
Remember to write all fractions in lowest terms.

1. 0.1 = ________ **2.** 0.7 = ________ **3.** 0.29 = ________

4. 0.03 = ________ **5.** 0.006 = ________ **6.** 0.072 = ________

7. 0.185 = ________ **8.** 0.503 = ________ **9.** 9.4 = ________

10. 2.851 = ________ **11.** 3.219 = ________ **12.** 17.26 = ________

Write each fraction as a decimal.

13. $\frac{5}{10}$ = ________ **14.** $\frac{1}{10}$ = ________ **15.** $\frac{7}{100}$ = ________

16. $\frac{42}{100}$ = ________ **17.** $\frac{706}{1000}$ = ________ **18.** $\frac{91}{1000}$ = ________

19. $4\frac{1}{10}$ = ________ **20.** $6\frac{38}{100}$ = ________ **21.** $9\frac{123}{1000}$ = ________

22. $1\frac{53}{1000}$ = ________ **23.** $7\frac{8}{100}$ = ________ **24.** $2\frac{2}{10}$ = ________

25. $\frac{3}{4}$ = ________ **26.** $\frac{1}{8}$ = ________ **27.** $\frac{7}{20}$ = ________

28. $\frac{125}{625}$ = ________ **29.** $\frac{19}{95}$ = ________ **30.** $\frac{27}{500}$ = ________

Use after pages 328–329.

SHARPEN YOUR SKILLS

Interpret the Remainder

Solve each problem.

1. Ann needed 150 hot dogs for her party. Hot dogs are sold by the pound. One pound contains 8 hot dogs. How many pounds should she buy?

   ____

2. Hot-dog buns are sold in packages of 12. How many packages should she buy for the 150 hot dogs? How many hot-dog buns will be left over?

   ____

3. Larry will bring the juice. One can of juice serves 4 people. How many cans are needed for the 75 people at the party?

   ____

4. Marie will bring the potato salad. One pound of potato salad serves 6 people. How many pounds are needed to serve 75 people?

   ____

5. Millie will bring baked beans. One dish of baked beans serves 12 people. How many dishes should she bring to serve the 75 people at the party?

   ____

6. Ice cream will be served for dessert. One quart of ice cream serves 7 people. How many quarts will be needed to serve 75 people?

   ____

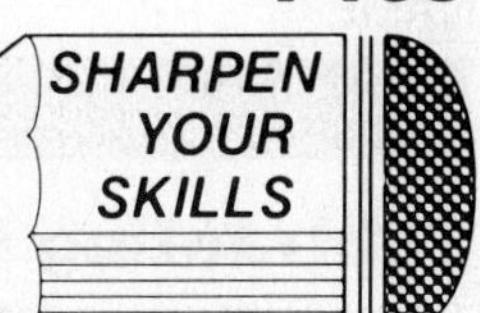

Adding Fractions: Same Denominator

Match each addition exercise with its answer. Draw a line from the star beside the exercise to the star beside the answer. Each line will go through a letter. Write the letter in the blank. You will spell the name of a famous baseball player.

1. $\frac{1}{7} + \frac{3}{7}$	★	★ $\frac{16}{24}$ or $\frac{2}{3}$	______
2. $\frac{7}{12} + \frac{5}{12}$	★	★ $\frac{6}{16}$ or $\frac{3}{8}$	______
3. $\frac{5}{9} + \frac{1}{9}$	★	★ $\frac{6}{9}$ or $\frac{2}{3}$	______
4. $\frac{5}{16} + \frac{1}{16}$	★	★ $1\frac{5}{10}$ or $1\frac{1}{2}$	______
5. $\frac{7}{20} + \frac{9}{20}$	★	★ 1	______
6. $\frac{7}{8} + \frac{3}{8}$	★	★ $1\frac{3}{6}$ or $1\frac{1}{2}$	______
7. $\frac{5}{9} + \frac{3}{9}$	★	★ $\frac{4}{7}$	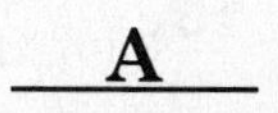 A
8. $\frac{7}{12} + \frac{11}{12}$	★	★ $\frac{5}{15}$ or $\frac{1}{3}$	______
9. $\frac{9}{10} + \frac{6}{10}$	★	★ $\frac{16}{20}$ or $\frac{4}{5}$	______
10. $\frac{7}{24} + \frac{9}{24}$	★	★ $\frac{8}{9}$	______
11. $\frac{5}{6} + \frac{4}{6}$	★	★ $1\frac{6}{12}$ or $1\frac{1}{2}$	______
12. $\frac{2}{15} + \frac{3}{15}$	★	★ $1\frac{2}{8}$ or $1\frac{1}{4}$	______

Letters: J, Y, E, T, E, S, A, H, V, L, O, T, C, R, N, B

NAME

SHARPEN YOUR SKILLS

Subtracting Fractions: Same Denominator

Subtract the fractions. **Remember** to simplify the answer if necessary.

1. $\frac{11}{12} - \frac{7}{12}$

2. $\frac{11}{12} - \frac{1}{12}$

3. $\frac{7}{10} - \frac{1}{10}$

4. $\frac{9}{20} - \frac{3}{20}$

5. $\frac{7}{10} - \frac{3}{10}$

6. $\frac{3}{4} - \frac{1}{4}$

7. $\frac{4}{5} - \frac{2}{5}$

8. $\frac{8}{9} - \frac{5}{9}$

9. $\frac{9}{10} - \frac{3}{10}$

10. $\frac{7}{9} - \frac{4}{9}$

11. $\frac{11}{12} - \frac{5}{12}$

12. $\frac{5}{6} - \frac{1}{6}$

13. $\frac{5}{8} - \frac{3}{8}$

14. $\frac{9}{20} - \frac{7}{20}$

15. $\frac{9}{10} - \frac{1}{10}$

16. $\frac{10}{12} - \frac{5}{12}$

Mixed Practice Add or subtract these fractions. Write your answer in simplest form.

17. $\frac{1}{6} + \frac{4}{6}$ ______

18. $\frac{7}{8} - \frac{3}{8}$ ______

19. $\frac{11}{9} - \frac{2}{9}$ ______

20. $\frac{3}{11} + \frac{5}{11}$ ______

21. $\frac{10}{12} - \frac{4}{12}$ ______

22. $\frac{3}{15} + \frac{10}{15}$ ______

23. $\frac{1}{4} + \frac{1}{4}$ ______

24. $\frac{6}{7} - \frac{3}{7}$ ______

25. $\frac{3}{9} + \frac{3}{9}$ ______

Use after pages 344–345.

NAME

SHARPEN YOUR SKILLS

Adding and Subtracting Mixed Numbers: Same Denominator

Add. **Remember** to write your answers in simplest form.

1. $8 + 7\frac{3}{5}$

2. $\frac{3}{10} + 6\frac{3}{10}$

3. $9\frac{3}{8} + \frac{2}{8}$

4. $6\frac{2}{9} + 7\frac{5}{9}$

5. $5\frac{3}{7} + 8\frac{2}{7}$

6. $8\frac{2}{5} + 9\frac{2}{5}$

7. $3\frac{3}{6} + 9\frac{2}{6}$

8. $6\frac{3}{8} + 9\frac{4}{8}$

9. $7\frac{1}{12} + 8\frac{4}{12} =$ ____________

10. $4\frac{3}{12} + 8\frac{3}{12} =$ ____________

11. $5\frac{3}{10} + 7\frac{5}{10} =$ ____________

12. $8\frac{3}{6} + 9\frac{1}{6} =$ ____________

Subtract. **Remember** to write your answers in simplest form.

13. $23\frac{7}{20} - 6\frac{3}{20}$

14. $14\frac{5}{24} - 5\frac{3}{24}$

15. $12\frac{4}{10} - 7\frac{1}{10}$

16. $17\frac{7}{16} - 9\frac{3}{16}$

17. $8\frac{14}{15} - 4\frac{7}{15}$

18. $17\frac{9}{12} - 9\frac{5}{12}$

19. $15\frac{7}{8} - 4\frac{3}{8}$

20. $15\frac{9}{10} - 11\frac{3}{10}$

21. $14\frac{7}{11} - 7\frac{4}{11}$

22. $12\frac{7}{9} - 8\frac{5}{9}$

23. $16\frac{5}{8} - 7\frac{4}{8}$

24. $7\frac{9}{20} - 4\frac{3}{20}$

NAME

SHARPEN YOUR SKILLS

Adding Mixed Numbers: Renaming Sums

Add. **Remember** to write your answers in simplest form.

1. $8\frac{4}{5} + \frac{2}{5}$

2. $\frac{1}{6} + 7\frac{5}{6}$

3. $12\frac{5}{7} + \frac{4}{7}$

4. $18\frac{3}{4} + \frac{3}{4}$

5. $5\frac{4}{9} + 4\frac{5}{9}$

6. $2\frac{2}{3} + 9\frac{2}{3}$

7. $4\frac{11}{12} + 8\frac{6}{12}$

8. $3\frac{6}{8} + 7\frac{5}{8}$

9. $7\frac{7}{10} + 8\frac{7}{10}$

10. $9\frac{3}{8} + 9\frac{7}{8}$

11. $9\frac{5}{9} + 6\frac{7}{9}$

12. $26\frac{7}{12} + 5\frac{11}{12}$

13. $38\frac{1}{6} + 9\frac{5}{6}$ ______

14. $15\frac{7}{12} + 8\frac{8}{12}$ ______

15. $23\frac{2}{9} + 4\frac{8}{9}$ ______

16. $41\frac{4}{5} + 14\frac{4}{5}$ ______

Solve the problem.

17. The stage crew made a backdrop for a play. They used $5\frac{3}{8}$ yards of white fabric and $2\frac{5}{8}$ yards of black fabric. How much fabric did they use?

______

NAME

SHARPEN YOUR SKILLS

Subtracting from a Whole Number

Subtract. **Remember** to write your answers in simplest form.

1. $1 - \frac{3}{5}$

2. $1 - \frac{7}{10}$

3. $1 - \frac{3}{4}$

4. $9 - 5\frac{4}{7}$

5. $7 - 3\frac{5}{6}$

6. $4 - 2\frac{8}{9}$

7. $8 - 7\frac{3}{4}$

8. $6 - 5\frac{7}{12}$

9. $10 - 4\frac{7}{8}$

10. $14 - 8\frac{2}{3}$

11. $17 - 8\frac{1}{8}$

12. $10 - 9\frac{7}{12}$

13. $34 - 11\frac{5}{8}$

14. $27 - 26\frac{3}{5}$

15. $19 - 12\frac{2}{9}$

16. $42 - 9\frac{11}{12}$

17. $31 - 8\frac{5}{8} =$ ______

18. $7 - \frac{2}{9} =$ ______

19. $9 - 8\frac{7}{10} =$ ______

20. Can you trace this figure without lifting your pencil from the paper and without retracing any line? Trace through your answers to the exercises in the order they are given.

____

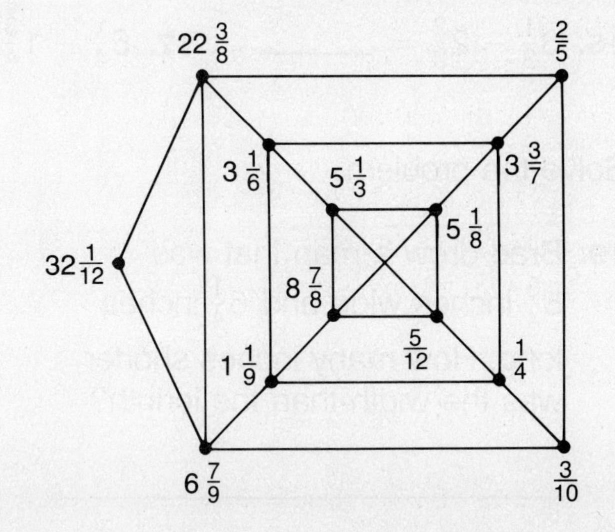

Use after pages 350–351.

NAME

Subtracting Mixed Numbers with Renaming

Subtract. **Remember** to write the answers in simplest form.

1. $9\frac{1}{3} - 5\frac{2}{3}$

2. $6\frac{3}{7} - 2\frac{6}{7}$

3. $8\frac{2}{5} - 6\frac{3}{5}$

4. $7\frac{1}{2} - 3\frac{1}{2}$

5. $7\frac{3}{8} - \frac{7}{8}$

6. $5\frac{3}{10} - \frac{1}{10}$

7. $2\frac{5}{9} - \frac{7}{9}$

8. $1\frac{1}{6} - \frac{5}{6}$

9. $9\frac{2}{5} - 8\frac{4}{5}$

10. $1\frac{1}{7} - \frac{3}{7}$

11. $8\frac{1}{10} - 4\frac{7}{10}$

12. $6\frac{5}{12} - 5\frac{11}{12}$

13. $32\frac{3}{8} - 8\frac{5}{8} =$ ______

14. $41\frac{7}{12} - 7\frac{5}{12} =$ ______

15. $22\frac{5}{9} - 16\frac{8}{9} =$ ______

16. $3\frac{1}{3} - 2\frac{2}{3} =$ ______

17. $6\frac{1}{4} - 1\frac{3}{4} =$ ______

18. $7\frac{1}{9} - 2\frac{5}{9} =$ ______

Solve the problem.

19. Brad drew a map that was $5\frac{3}{4}$ inches wide and $6\frac{1}{4}$ inches long. How many inches shorter was the width than the length?

____

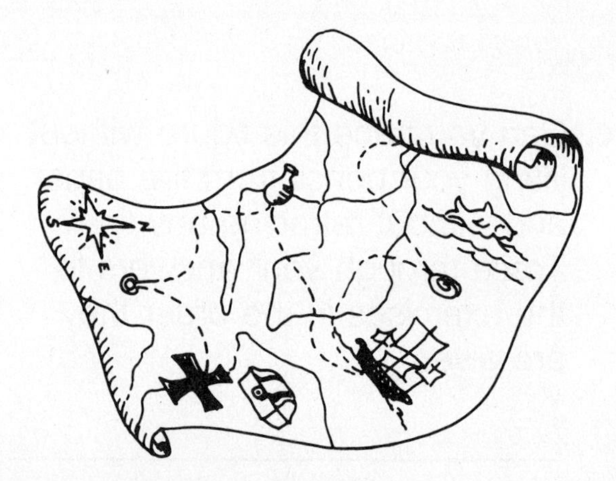

NAME

SHARPEN YOUR SKILLS

Adding and Subtracting Fractions: Related Denominators

Add or subtract by finding a common denominator.
Remember to write your answers in simplest form.

1. $\frac{5}{20} + \frac{3}{10}$

2. $\frac{19}{24} - \frac{5}{12}$

3. $\frac{1}{2} + \frac{5}{8}$

4. $\frac{7}{9} - \frac{1}{3}$

5. $\frac{11}{12} - \frac{5}{6}$

6. $\frac{7}{8} + \frac{3}{16}$

7. $\frac{7}{12} - \frac{1}{4}$

8. $\frac{14}{15} - \frac{2}{5}$

9. $\frac{5}{8} + \frac{1}{2}$

10. $\frac{31}{32} - \frac{5}{16}$

11. $\frac{15}{16} - \frac{1}{4}$

12. $\frac{5}{6} + \frac{2}{3}$

13. $\frac{11}{15} - \frac{2}{5} =$ ______

14. $\frac{9}{10} + \frac{1}{5} =$ ______

15. $\frac{11}{16} - \frac{1}{8} =$ ______

16. $\frac{9}{10} + \frac{2}{5} =$ ______

Mixed Practice Add or subtract.

17. $7\frac{7}{9} - 5\frac{2}{9} =$ ______

18. $1\frac{2}{5} + 1\frac{1}{5} =$ ______

19. $11\frac{1}{4} - 8\frac{1}{4} =$ ______

20. $7 - 1\frac{2}{3} =$ ______

21. $12 - 4\frac{5}{6} =$ ______

22. $19\frac{1}{6} - 13\frac{5}{6} =$ ______

Use after pages 356–357.

SHARPEN YOUR SKILLS

Use Alternate Strategies

In the space on the right, draw a diagram to solve each problem. Then use an alternate strategy.

1. The library is building new book shelves. One set will go on a wall 10 feet wide and $12\frac{1}{2}$ feet tall. The book shelves will be 6 feet wide. The librarian needs $3\frac{1}{2}$ feet on one side to put a chair. How much room will be left on the other side?

____

____

2. The librarian will put 40 books on each shelf. Each book is $1\frac{1}{2}$ inches thick. How much space will be left on each shelf to add new books?

____

____

3. The dictionary shelf will have books that are 3 inches wide. How many dictionaries will fit on the shelf? How many shelves will be needed to store 72 books?

____

____

____

Use after pages 358–359.

NAME

SHARPEN YOUR SKILLS

Adding Fractions: Different Denominators

Find each sum. **Remember** to write your answers in simplest form.

1. $\frac{1}{2} + \frac{1}{5}$

2. $\frac{3}{10} + \frac{2}{15}$

3. $\frac{1}{8} + \frac{5}{12}$

4. $\frac{3}{9} + \frac{1}{6}$

5. $\frac{2}{5} + \frac{3}{4}$

6. $\frac{1}{8} + \frac{5}{6}$

7. $\frac{3}{4} + \frac{3}{5}$

8. $\frac{7}{12} + \frac{1}{8}$

9. $\frac{5}{8} + \frac{5}{6} =$ ______

10. $\frac{1}{4} + \frac{9}{10} =$ ______

11. $\frac{3}{4} + \frac{1}{6} =$ ______

12. $\frac{2}{3} + \frac{3}{8} =$ ______

13. $\frac{1}{3}$, $\frac{3}{4}$, $+ \frac{5}{6}$

14. $\frac{1}{6}$, $\frac{2}{3}$, $+ \frac{5}{9}$

15. $\frac{2}{5}$, $\frac{7}{10}$, $+ \frac{3}{4}$

16. $\frac{4}{6}$, $\frac{2}{3}$, $+ \frac{1}{5}$

Solve the problem.

17. If $\frac{2}{3}$ of a book were maps and another $\frac{1}{5}$ were pictures, what fraction of the book would be pictures and maps together?

______

Use after pages 360–363.

SHARPEN YOUR SKILLS

Subtracting Fractions: Different Denominators

Find each difference. **Remember** to write your answers in simplest form.

1. $\frac{5}{6} - \frac{1}{4}$

2. $\frac{3}{4} - \frac{1}{3}$

3. $\frac{5}{6} - \frac{5}{9}$

4. $\frac{7}{10} - \frac{3}{15}$

5. $\frac{1}{2} - \frac{2}{5}$

6. $\frac{5}{6} - \frac{1}{9}$

7. $\frac{9}{10} - \frac{3}{4}$

8. $\frac{5}{8} - \frac{1}{6}$

9. $\frac{4}{7} - \frac{1}{4}$

10. $\frac{9}{15} - \frac{3}{10}$

11. $\frac{4}{5} - \frac{3}{4}$

12. $\frac{11}{12} - \frac{3}{4}$

13. $\frac{7}{8} - \frac{2}{5} =$ ______

14. $\frac{2}{6} - \frac{2}{9} =$ ______

15. $\frac{4}{5} - \frac{1}{4} =$ ______

16. $\frac{7}{8} - \frac{1}{3} =$ ______

Solve the problem.

17. Sam used $\frac{3}{4}$ cup oat flour and $\frac{1}{6}$ cup honey in a recipe. How much more oat flour than honey did he use?

______

NAME

SHARPEN YOUR SKILLS

Adding Mixed Numbers: Different Denominators

Add. **Remember** to write your answers in simplest form.

1. $7\frac{2}{3} + \frac{1}{12}$

2. $9\frac{2}{5} + \frac{1}{4}$

3. $6\frac{2}{5} + 5\frac{3}{10}$

4. $3\frac{3}{5} + 9\frac{1}{3}$

5. $9\frac{5}{6} + 5\frac{2}{3}$

6. $6\frac{7}{15} + 8\frac{1}{3}$

7. $9\frac{3}{4} + 8\frac{2}{3}$

8. $9\frac{5}{6} + 9\frac{5}{12}$

9. $1\frac{3}{5} + 3\frac{3}{4} + 4\frac{7}{20}$

10. $1\frac{2}{3} + 26\frac{5}{6} + 1\frac{7}{12}$

11. $14\frac{1}{4} + 52\frac{3}{8} + 12\frac{9}{16}$

12. $11\frac{3}{4} + 27\frac{7}{10} + 31\frac{4}{5}$

13. $11\frac{7}{12} + \frac{1}{6} =$ ______

14. $32\frac{3}{8} + 12\frac{1}{4} =$ ______

15. $41\frac{3}{4} + 16\frac{3}{5} =$ ______

16. $53\frac{1}{2} + 14\frac{3}{5} =$ ______

Mental Math Find each sum mentally.

17. $4\frac{1}{4} + 2\frac{2}{3} + 3\frac{3}{4} =$ ______

18. $9\frac{1}{8} + 3\frac{3}{5} + 2\frac{2}{5} =$ ______

19. $8\frac{5}{6} + 4\frac{5}{7} + 7\frac{1}{6} =$ ______

20. $3\frac{4}{9} + 3\frac{5}{9} + 4\frac{4}{5} =$ ______

21. $5\frac{11}{15} + 3\frac{4}{15} + 5\frac{3}{8} =$ ______

22. $5\frac{2}{3} + 4\frac{7}{10} + 6\frac{3}{10} =$ ______

Use after pages 366–367.

SHARPEN YOUR SKILLS

Subtracting Mixed Numbers: Different Denominators

Subtract. **Remember** to write your answers in simplest form.

1. $7\frac{5}{8} - \frac{1}{2}$

2. $16\frac{1}{12} - 9\frac{3}{4}$

3. $4\frac{7}{10} - 4\frac{2}{5}$

4. $8\frac{2}{3} - 3\frac{1}{6}$

5. $12\frac{1}{4} - 5\frac{7}{8}$

6. $11\frac{1}{10} - \frac{1}{2}$

7. $15\frac{5}{6} - 6\frac{1}{2}$

8. $19\frac{11}{12} - 4\frac{2}{3}$

9. $18\frac{2}{5} - 9\frac{1}{2}$

10. $6\frac{3}{4} - 6\frac{2}{3}$

11. $13\frac{1}{3} - 3\frac{3}{5}$

12. $14\frac{2}{3} - 8\frac{1}{4}$

13. $11\frac{5}{8} - 8\frac{1}{3} =$ ______

14. $27\frac{1}{4} - 6\frac{2}{5} =$ ______

Use the table to find the differences in average annual rainfall for each pair of cities.

City	Average annual rainfall
Montgomery, AL	$49\frac{1}{8}$ in.
Phoenix, AZ	$7\frac{1}{10}$
Los Angeles, CA	$12\frac{1}{10}$
Miami, FL	$57\frac{1}{2}$
Atlanta, GA	$48\frac{3}{5}$
Jackson, MS	$52\frac{7}{8}$
Albuquerque, NM	$8\frac{1}{6}$
Houston, TX	$44\frac{3}{4}$

15. Montgomery and Jackson ______ in.

16. Miami and Atlanta ______ in.

17. Los Angeles and Phoenix ______ in.

18. Jackson and Albuquerque ______ in.

NAME

SHARPEN YOUR SKILLS

Solve a Simpler Problem

Rob plans the travel for sightseeing trips. Show what simpler numbers you could use to solve each problem. Then solve the problem.

1. Travel from Fun Travel to Rivergate Park will take $1\frac{3}{4}$ hours. It will take $1\frac{1}{4}$ hours to go from the park to the Kennedy Memorial. Then it will take $1\frac{1}{2}$ hours to return to Fun Travel. How long will the trip be?

____

____

____

2. Rob plans a walking tour of River City. The guide leads the group $\frac{6}{10}$ mile from Fun Travel to the seaport. Then they walk $\frac{1}{12}$ mile to a cafe. After lunch they walk $1\frac{1}{4}$ mile across the bridge. They return to Fun Travel by bus. How far does the group walk?

____

____

____

3. The children's trip goes to Play World. It takes $\frac{3}{4}$ hour to get to Play World. They spend $1\frac{1}{6}$ hours there. The bus ride back to Fun Travel is $\frac{1}{2}$ hour. How long will the trip be?

____

____

____

4. A science trip visits the Space Museum, the Planetarium, and the Space Ride. It is $10\frac{1}{8}$ miles from Fun Travel to the Space Museum, $8\frac{3}{4}$ miles to the Planetarium, and $7\frac{1}{2}$ miles to the Space Ride. How many miles will the trip cover one way?

____

____

____

SHARPEN YOUR SKILLS

Using Pictures to Multiply Fractions

Use each picture to find the product.

1. $\frac{1}{4}$ of $\frac{2}{3}$

$\frac{1}{4} \times \frac{2}{3} =$ ______

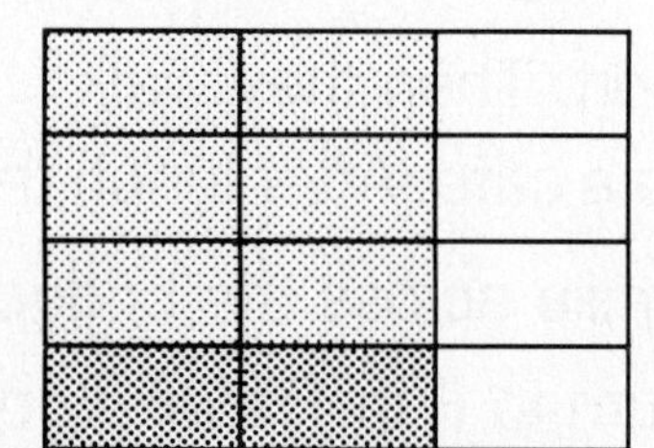

2. $\frac{2}{3}$ of $\frac{2}{3}$

$\frac{2}{3} \times \frac{2}{3} =$ ______

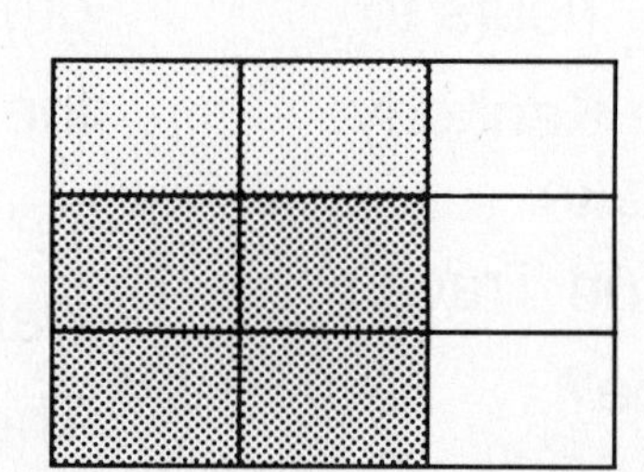

3. $\frac{2}{3}$ of $\frac{3}{5}$

$\frac{2}{3} \times \frac{3}{5} =$ ______

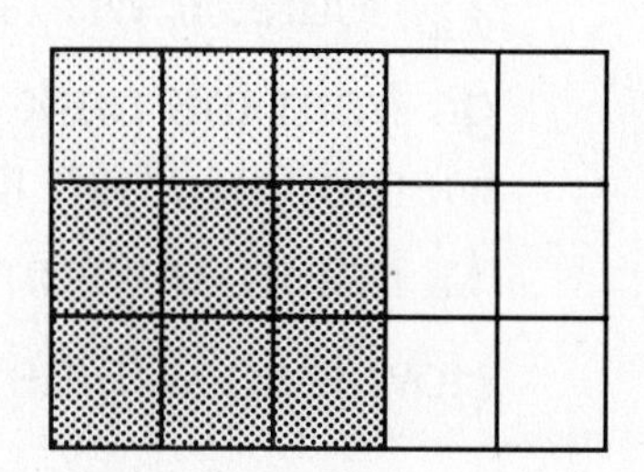

Write a number sentence that states what each picture shows.

4.

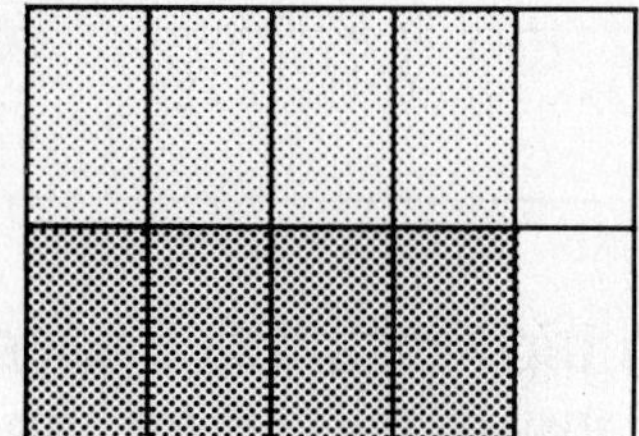

______

5.

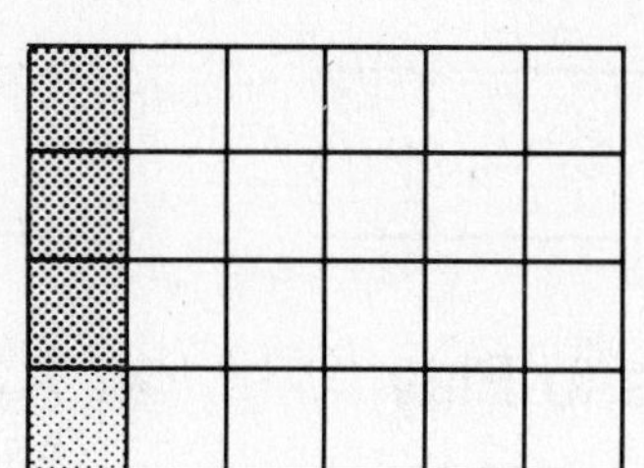

______

6.

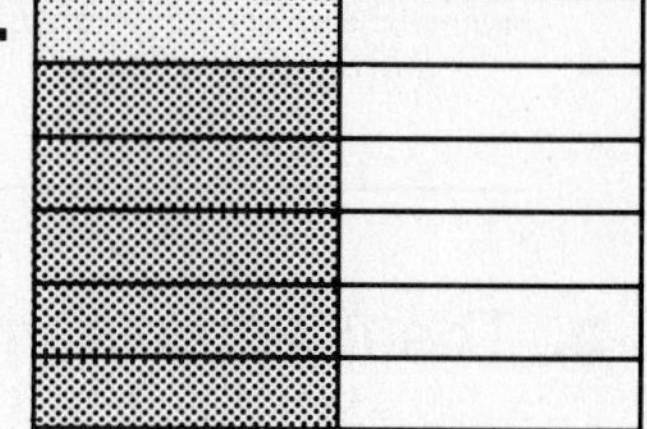

______

Draw a picture to find each product.

7. $\frac{1}{3}$ of $\frac{4}{5}$

$\frac{1}{3} \times \frac{4}{5} =$ ______

8. $\frac{1}{2}$ of $\frac{2}{3}$

$\frac{1}{2} \times \frac{2}{3} =$ ______

9. $\frac{3}{5}$ of $\frac{3}{4}$

$\frac{3}{5} \times \frac{3}{4} =$ ______

SHARPEN YOUR SKILLS

Multiplying Fractions

Find the height of the world's highest peak, Mount Everest. First, multiply to find the answer to each problem. **Remember** to write each product in simplest form. Then cross out the digit over each answer in the chart below. The remaining digits give the height of Mount Everest in feet.

1. $\frac{1}{2} \times \frac{3}{4} =$ ________ **2.** $\frac{1}{6} \times \frac{1}{3} =$ ________ **3.** $\frac{3}{4} \times \frac{3}{5} =$ ________

4. $\frac{2}{3} \times \frac{1}{2} =$ ________ **5.** $\frac{5}{7} \times \frac{1}{2} =$ ________ **6.** $\frac{2}{5} \times \frac{2}{9} =$ ________

7. $\frac{1}{5} \times \frac{2}{3} =$ ________ **8.** $\frac{3}{4} \times \frac{4}{5} =$ ________ **9.** $\frac{2}{3} \times \frac{6}{7} =$ ________

10. $\frac{4}{5} \times \frac{2}{7} =$ ________ **11.** $\frac{8}{9} \times \frac{3}{4} =$ ________ **12.** $\frac{6}{7} \times \frac{3}{4} =$ ________

13. $\frac{1}{3} \times \frac{10}{11} =$ ________ **14.** $\frac{7}{8} \times \frac{2}{5} =$ ________ **15.** $\frac{1}{3} \times \frac{2}{3} =$ ________

16. $\frac{1}{2} \times \frac{1}{3} \times \frac{1}{5} =$ ____________ **17.** $\frac{1}{3} \times \frac{1}{5} \times \frac{2}{3} =$ ____________

3	2	1	0	8	6	9	4	5	7	2	0	3	9	1	8	4	2	8	5	3	7
$\frac{2}{45}$	$\frac{16}{25}$	$\frac{3}{8}$	$\frac{8}{35}$	$\frac{5}{14}$	$\frac{10}{33}$	$\frac{9}{10}$	$\frac{7}{20}$	$\frac{9}{20}$	$\frac{1}{30}$	$\frac{3}{5}$	$\frac{1}{50}$	$\frac{2}{9}$	$\frac{1}{18}$	$\frac{2}{3}$	$\frac{2}{15}$	$\frac{4}{7}$	$\frac{7}{11}$	$\frac{8}{15}$	$\frac{1}{3}$	$\frac{9}{14}$	$\frac{4}{45}$

NAME

SHARPEN YOUR SKILLS

Multiplying Fractions and Whole Numbers

Each frame holds a number.

┐ holds the number 10.

└ holds the number 6.

Write the correct numbers in the frames below. Then find each product. **Remember** to write the product in simplest form.

12	$\frac{3}{4}$	6
$\frac{5}{6}$	$\frac{2}{3}$	$\frac{7}{10}$
10	$\frac{3}{5}$	15

1. $\frac{7}{10}$ × 10 = ______

2. ____ × ____ = ______

3. ____ × ____ = ______

4. ____ × ____ = ______

5. ____ × ____ = ______

6. ____ × ____ = ______

7. ____ × ____ = ______

8. ____ × ____ = ______

9. ____ × ____ = ______

10. ____ × ____ = ______

Multiply.

11. $\frac{1}{4} \times 12 =$ ______ **12.** $\frac{4}{5} \times 25 =$ ______ **13.** $\frac{2}{3} \times 15 =$ ______

Use after pages 392–393.

SHARPEN YOUR SKILLS

Mixed Numbers as Improper Fractions

Write each mixed number as an improper fraction.

1. $3\frac{1}{3}$ = ______ **2.** $50\frac{1}{2}$ = ______ **3.** $1\frac{7}{8}$ = ______

4. $12\frac{5}{16}$ = ______ **5.** $9\frac{7}{18}$ = ______ **6.** $14\frac{2}{7}$ = ______

7. $16\frac{9}{10}$ = ______ **8.** $7\frac{6}{7}$ = ______ **9.** $90\frac{7}{12}$ = ______

10. $15\frac{2}{3}$ = ______ **11.** $22\frac{3}{8}$ = ______ **12.** $19\frac{5}{12}$ = ______

Mixed Practice Write a mixed number or a whole number for each fraction.

13. $\frac{12}{3}$ = ______ **14.** $\frac{18}{4}$ = ______ **15.** $\frac{36}{5}$ = ______

16. $\frac{92}{12}$ = ______ **17.** $\frac{53}{7}$ = ______ **18.** $\frac{41}{6}$ = ______

19. $\frac{14}{14}$ = ______ **20.** $\frac{6}{2}$ = ______ **21.** $\frac{63}{8}$ = ______

22. $\frac{115}{15}$ = ______ **23.** $\frac{12}{7}$ = ______ **24.** $\frac{100}{45}$ = ______

Solve the problem.

25. Janet is making a recipe that calls for $5\frac{1}{2}$ cups of juice. Her measuring cup holds only $\frac{1}{2}$ cup. Write the mixed number as an improper fraction. Does the numerator or the denominator tell you how many $\frac{1}{2}$ cup measures she needs?

____

NAME

SHARPEN YOUR SKILLS

Multiplying Mixed Numbers

What has four eyes but cannot see?

To answer this riddle, multiply to solve each problem. Write the letter of each product in the correct blank below. Some letters are not used.

1. $\frac{1}{2} \times 4\frac{1}{8} =$ ____ **I**

2. $2\frac{1}{3} \times 5 =$ ____ **P**

3. $5\frac{1}{4} \times 1\frac{1}{2} =$ ____ **I**

4. $4\frac{1}{10} \times \frac{1}{3} =$ ____ **H**

5. $4\frac{3}{4} \times \frac{3}{4} =$ ____ **M**

6. $2\frac{2}{3} \times 2\frac{1}{4} =$ ____ **N**

7. $1\frac{4}{5} \times 1\frac{1}{3} =$ ____ **S**

8. $7 \times 3\frac{1}{6} =$ ____ **R**

9. $4\frac{1}{2} \times \frac{5}{6} =$ ____ **I**

10. $2\frac{3}{4} \times 3\frac{1}{2} =$ ____ **P**

11. $5 \times 2\frac{2}{5} =$ ____ **S**

12. $1\frac{3}{5} \times 2\frac{1}{2} =$ ____ **T**

13. $\frac{4}{5} \times 2\frac{1}{8} =$ ____ **S**

14. $1\frac{3}{4} \times 9 =$ ____ **E**

15. $6 \times 2\frac{1}{4} =$ ____ **G**

16. $3\frac{1}{3} \times \frac{3}{4} =$ ____ **A**

17. $3\frac{1}{2} \times 1\frac{1}{2} =$ ____ **S**

18. $\frac{2}{3} \times 5\frac{1}{2} =$ ____ **I**

____ ____ ____

4 $1\frac{11}{30}$ $15\frac{3}{4}$

____ ____ ____ ____ ____ ____ ____ ____ ____ ____ ____

$3\frac{9}{16}$ $7\frac{7}{8}$ $2\frac{2}{5}$ 12 $2\frac{1}{16}$ $1\frac{7}{10}$ $5\frac{1}{4}$ $3\frac{3}{4}$ $11\frac{2}{3}$ $9\frac{5}{8}$ $3\frac{2}{3}$

Use after pages 398–399.

NAME ____________

SHARPEN YOUR SKILLS

Use Data from a Table

Use data from the table to solve each problem.

Length		Weight		Time	
1 foot	= 12 inches	1 pound	= 16 ounces	1 minute	= 60 seconds
1 yard	= 3 feet	1 ton	= 2,000 pounds	1 hour	= 60 minutes
1 mile	= 5,280 feet			1 day	= 24 hours

1. Eric worked $3\frac{1}{2}$ hours a day on the garden. He worked for 8 days. How many hours did he work? How many minutes is this?

____

2. The garden fence is $5\frac{1}{2}$ feet high. How many inches is this?

____

3. In his garden, Eric planted 1 row of pole beans that was $3\frac{1}{2}$ yards long. There are 24 plants in each yard. How many pole bean plants did Eric plant in the garden?

____

4. One garden plant grew $\frac{1}{2}$ inch each day. How much would the plant grow in a week? Another plant grew 1 inch per day. How much taller than the first plant would it be after 1 week?

____

5. Eric wanted to cover 300 feet of paths with gravel. It took $6\frac{2}{3}$ pounds of gravel to cover each foot of the path. How many pounds of gravel did he need? How many tons of gravel is this?

____

6. Eric's brother Josh put a bird feeder in the garden. The feeder held about $\frac{1}{2}$ pound of birdseed. They filled the feeder 50 times from one bag. How many pounds did the bag hold? How many ounces is that?

____

7. Eric and Josh read that a garden snail can crawl $\frac{1}{25}$ mile each hour. How many feet can the snail crawl in an hour?

____

Use after pages 400–401.

NAME

SHARPEN YOUR SKILLS

Exploring Division by Fractions

Use the pictures to complete the divisions.

1. 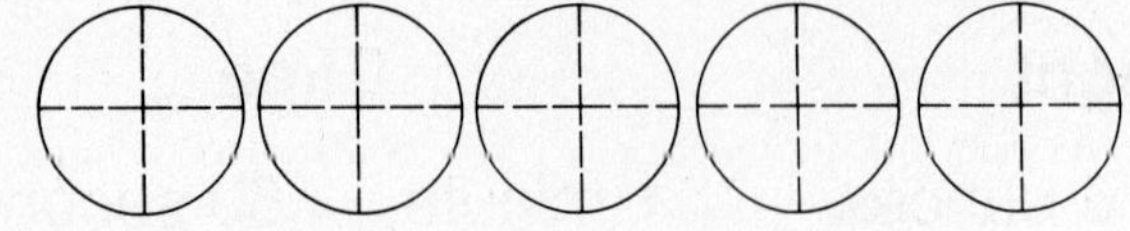

How many $\frac{1}{4}$s are in 5?

$5 \div \frac{1}{4} =$ ____________

2. 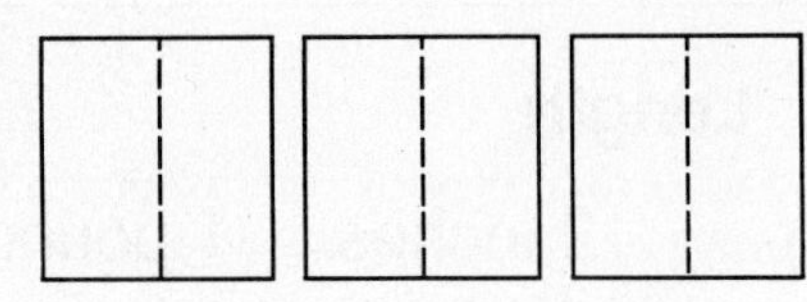

How many $\frac{1}{2}$s are in 3?

$3 \div \frac{1}{2} =$ ____________

3. 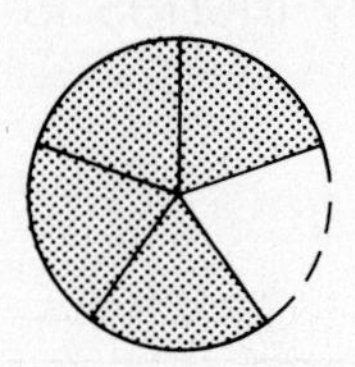

How many $\frac{1}{5}$s are in $\frac{4}{5}$?

$\frac{4}{5} \div \frac{1}{5} =$ ____________

4. 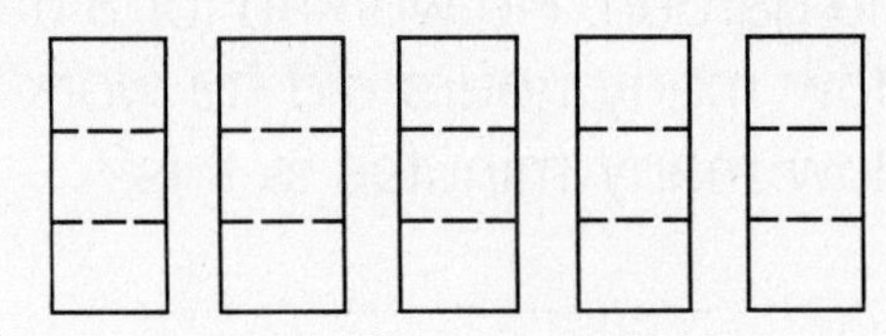

How many $\frac{1}{3}$s are in 5?

$5 \div \frac{1}{3} =$ ____________

5. 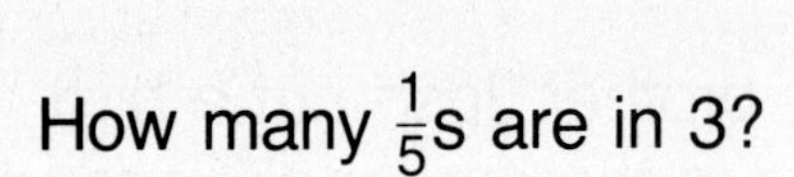

How many $\frac{1}{5}$s are in 3?

$3 \div \frac{1}{5} =$ ____________

6. 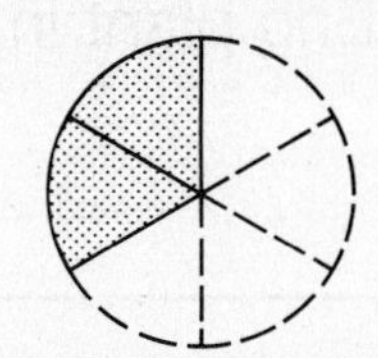

How many $\frac{1}{6}$s are in $\frac{1}{3}$?

$\frac{1}{3} \div \frac{1}{6} =$ ____________

Write a division sentence for each picture.

7. 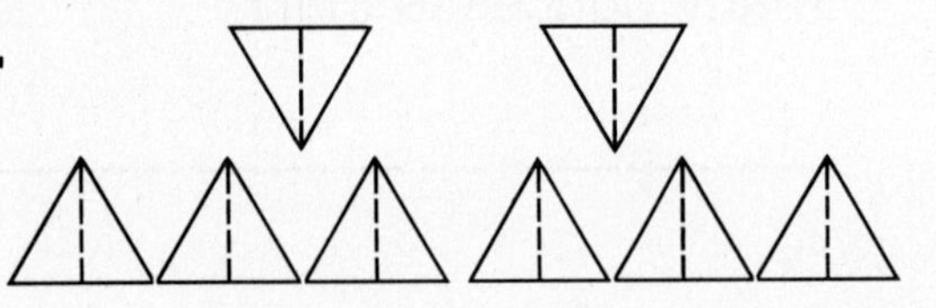

____

8. 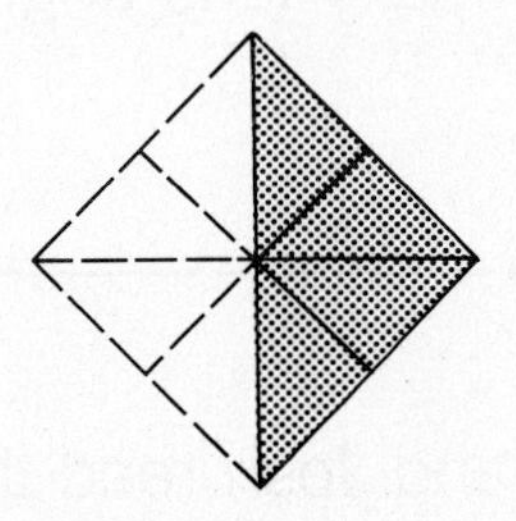

____

NAME ____________________

SHARPEN YOUR SKILLS

Choose an Operation

Solve each problem.

1. Ms. Adamson bought a doll house as a birthday gift. The house was marked $\frac{1}{4}$ off the original price of $68.20. How much did she pay for the doll house?

____

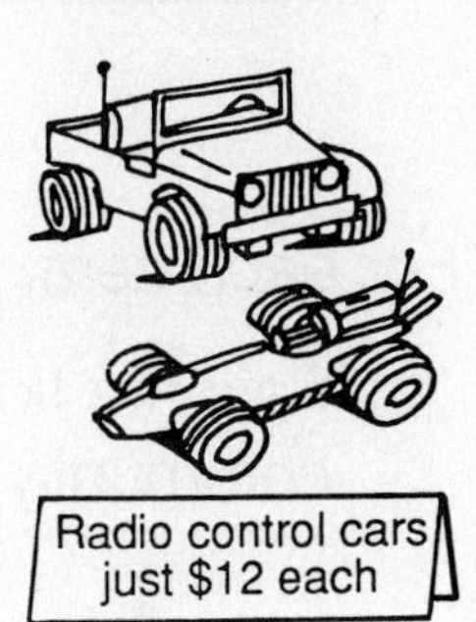

2. Radio control cars are on sale for $12 each. A week earlier, the same cars cost $18 each. How much is the discount? What fraction of the original price do the cars cost on sale?

____

3. A computer desk and chair set for Coretta's room was advertised at $\frac{1}{4}$ off the original price of $364. A regular desk was on sale for $\frac{1}{3}$ off the original price of $345. The matching chair was only $25. Which would be the better bargain for Coretta, the computer desk and chair or the regular desk and chair?

____

Critical Thinking Arnette wanted a 24-inch bike. She looked at a $120 bike marked $\frac{1}{4}$ off. Besides the discount, the store was giving a rebate of $10. The sales tax was $0.07 on a dollar. In another store, a much better bike was advertised at its regular price of $98, plus the $0.07 sales tax. Which bike would cost more? How much more? Which bike do you think Arnette should buy?

____

Use after pages 404–405.

NAME

SHARPEN YOUR SKILLS

Collecting Data

1. When collecting data, why would you survey only part of the population?

2. If your sample population is too small, how would that affect the survey?

For each item, tell who you would survey and the question you would ask.

3. Favorite flower of people in your community

4. Favorite extracurricular activity of students in your school

5. Favorite animal of students in your school

6. Favorite sports activity of students in your school

7. Number of hours per week spent watching television by elementary students nationwide

8. Number of fifth graders in your state who eat breakfast before going to school

Use after pages 416–417.

NAME

SHARPEN YOUR SKILLS

Organizing Data

Number of Rainy Days in Spring

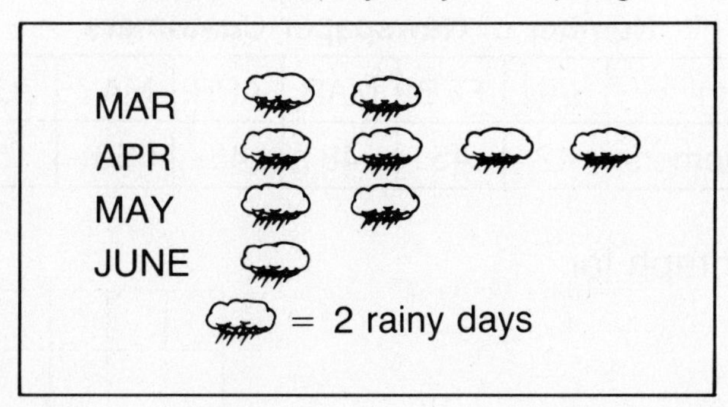

1. Which month had the most rain?

____

2. Which month had the least rain?

____

3. How many more rainy days were there in April than in June?

____

4. What kind of graph would best show how spring rainfall has increased or decreased over the last 10 years?

____

5. The teachers at Eastmont school met to vote on the time of day they wanted for their lunch hour. 5 chose 1:00, 15 chose 12:30, 10 chose 12:00, and 25 chose 11:30. Organize the data in the table to the right.

Preferred lunch hour				
Number of Teachers				

6. What kind of graph would you use to show the data. Why?

____

____

7. Make a graph in the box to the right to show the data.

Use after pages 418–421.

NAME

SHARPEN YOUR SKILLS

Making a Broken-Line Graph

DATA

Number of Newspaper Customers						
Month	JAN	FEB	MAR	APR	MAY	JUNE
Customers	42	45	48	45	50	53

Make a broken-line graph for the given data.

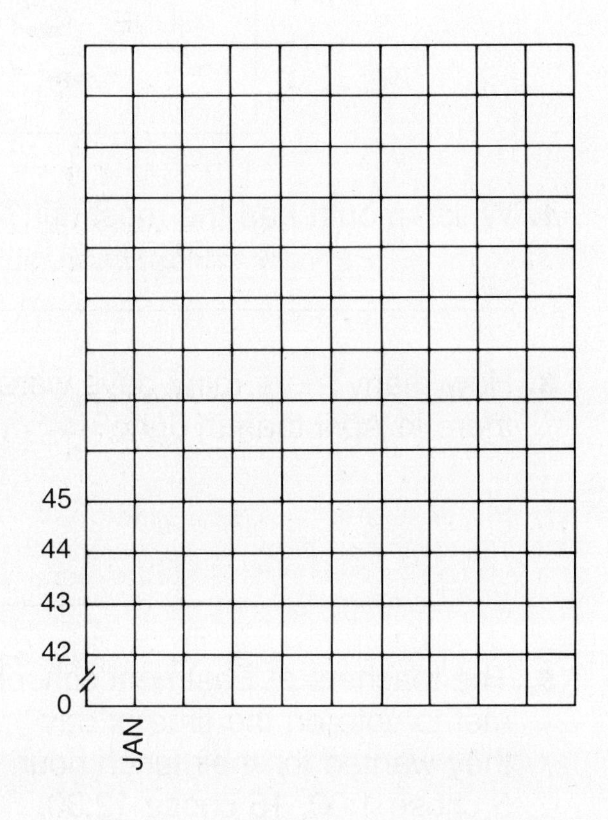

1. Finish setting up the vertical and horizontal scales, label them, and write a title.

2. Plot the data points on the graph and connect them.

3. What is the size of your first interval on the vertical scale?

___

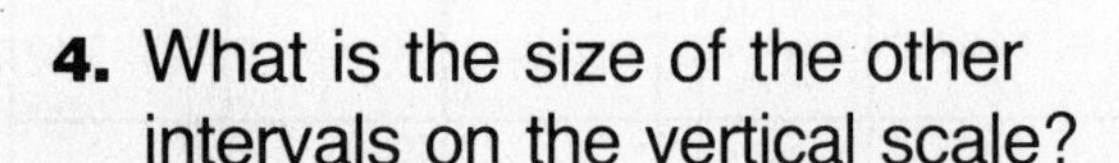

4. What is the size of the other intervals on the vertical scale?

___

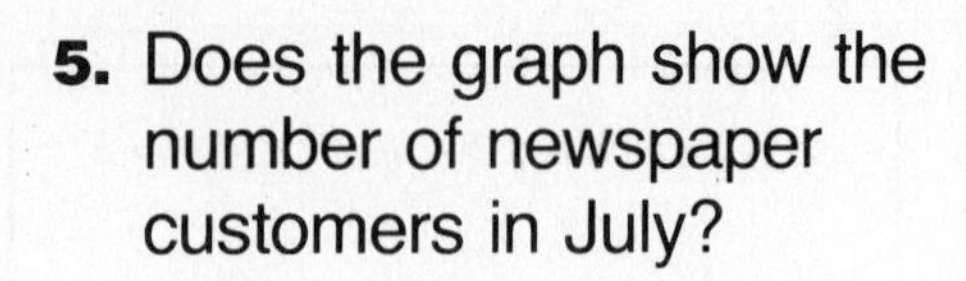

5. Does the graph show the number of newspaper customers in July?

___

6. Was the number of customers the same between any 2 months in a row?

___

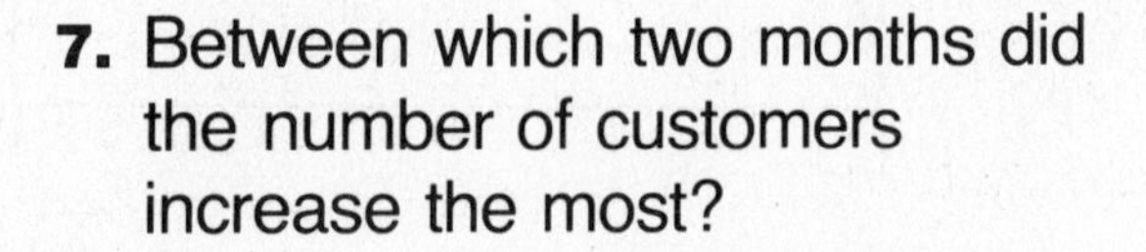

7. Between which two months did the number of customers increase the most?

___

8. Was the increase in the number of customers the same between any two months? What months?

___

___

NAME

Reading and Interpreting a Line Graph

This line graph shows how many practice problems David reviewed when he studied for his math test. Use this graph to solve each problem.

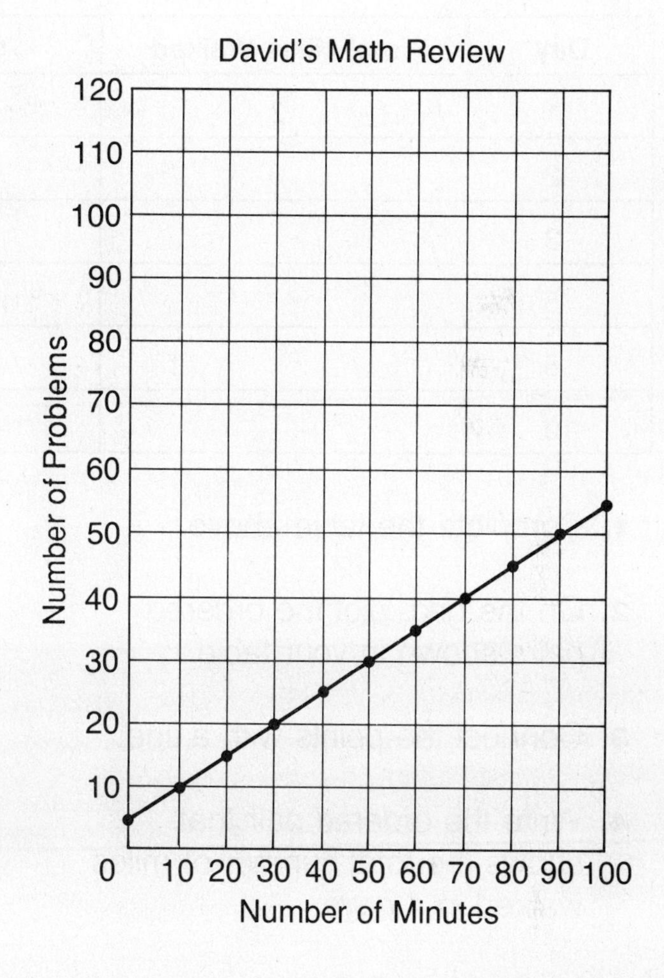

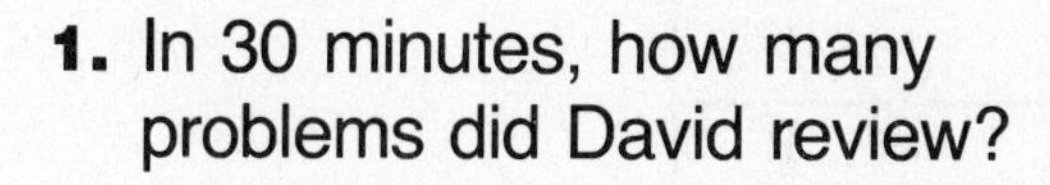

1. In 30 minutes, how many problems did David review?

____

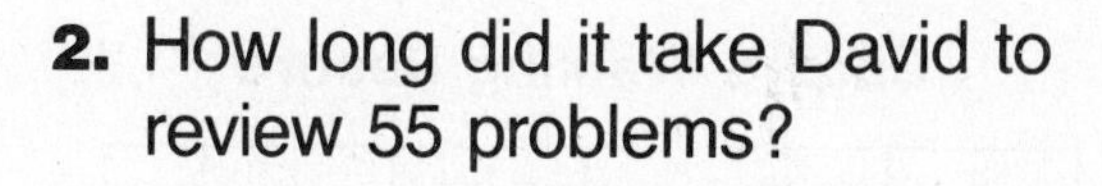

2. How long did it take David to review 55 problems?

____

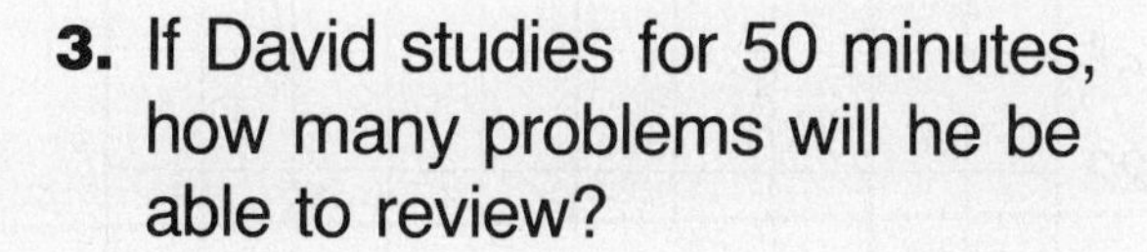

3. If David studies for 50 minutes, how many problems will he be able to review?

____

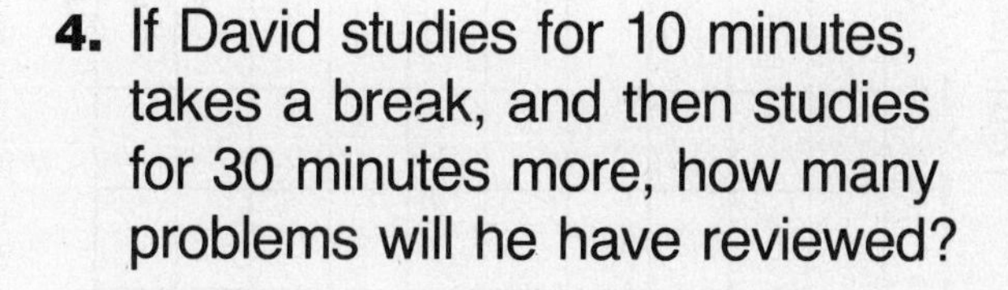

4. If David studies for 10 minutes, takes a break, and then studies for 30 minutes more, how many problems will he have reviewed?

____

5. How many more problems can David review in 50 minutes than in 40 minutes?

____

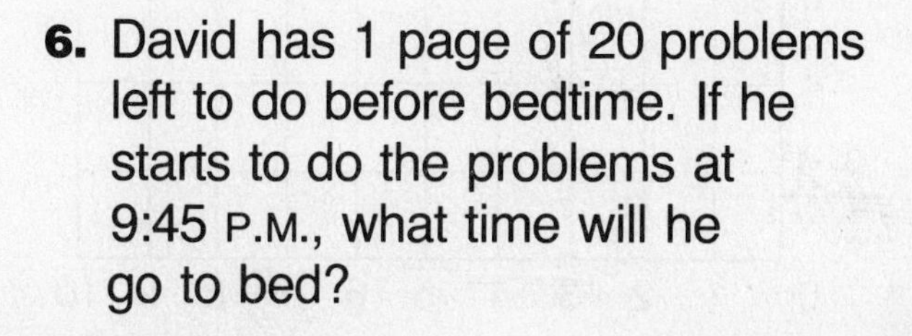

6. David has 1 page of 20 problems left to do before bedtime. If he starts to do the problems at 9:45 P.M., what time will he go to bed?

____

7. David left the last 50 problems to do in the morning before he leaves for school at 8:30 A.M. At what time must David start his work?

____

NAME

SHARPEN YOUR SKILLS

Making a Line Graph

Carol walks 3 miles every day.

Day	Total Miles Walked	Ordered Pair
1	3	(1, 3)
2	6	(2, 6)
3		
5		
8		
10		

1. Complete the table above.

2. On the grid, plot the ordered pairs shown in your table.

3. Connect the points with a line.

4. Write the ordered pair that shows the total number of miles walked in 4 days.

____

Use your line graph to answer each question.

5. How many miles did Carol walk in 6 days?

____

6. How many days does it take Carol to walk 21 miles?

____

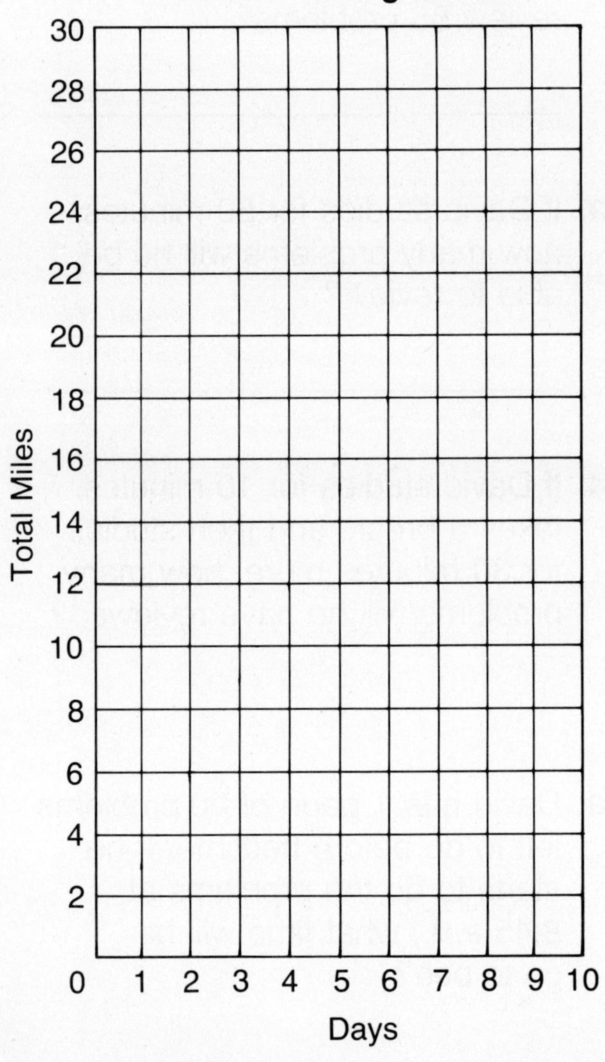

NAME

SHARPEN YOUR SKILLS

Make a Graph

What kind of graph would you use for each situation? Why?

1. Total sales for the top 5 albums for the month of June

2. The annual increase in VCR sales from 1980 to 1989

3. The number of women's snowboots sold for the month of January in 5 cities

4. The number of visitors to the school band performances on Wednesday, Thursday, Friday, and Saturday

5. Favorite subjects taken by a fifth-grade class at the Rollingwood School

6. The number of hours John worked at his after-school job during the month of June

7. The homerun record for Hank Aaron during his baseball career

8. The growth in population of suburban areas from 1965 to 1989

Answer each question.

9. During one week in March, it snowed 1 inch on the first day, 2 inches on the second day, 1.5 inches on the third day, 6 inches on the fourth day, and 2 inches on the fifth day. Use this data to make a graph to the right.

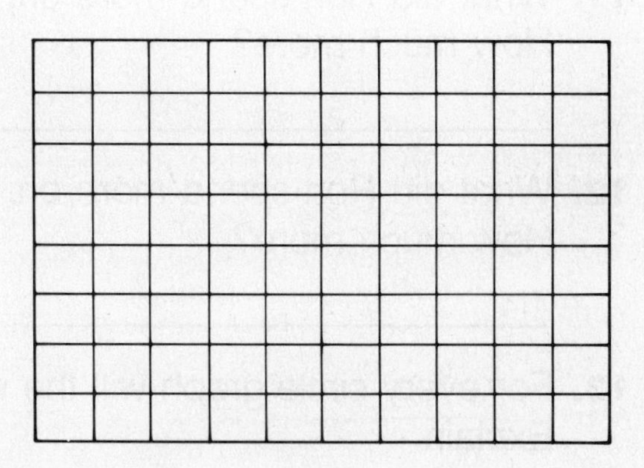

10. What kind of graph did you make to show the data in Problem 9? Why?

NAME

SHARPEN YOUR SKILLS

Circle Graphs

This circle graph shows how Ron spent $120 of his earnings.

What fraction of the $120 was spent on each?

1. Lunches ______
2. Movies ______
3. Clothing ______
4. Gifts ______
5. Snacks ______

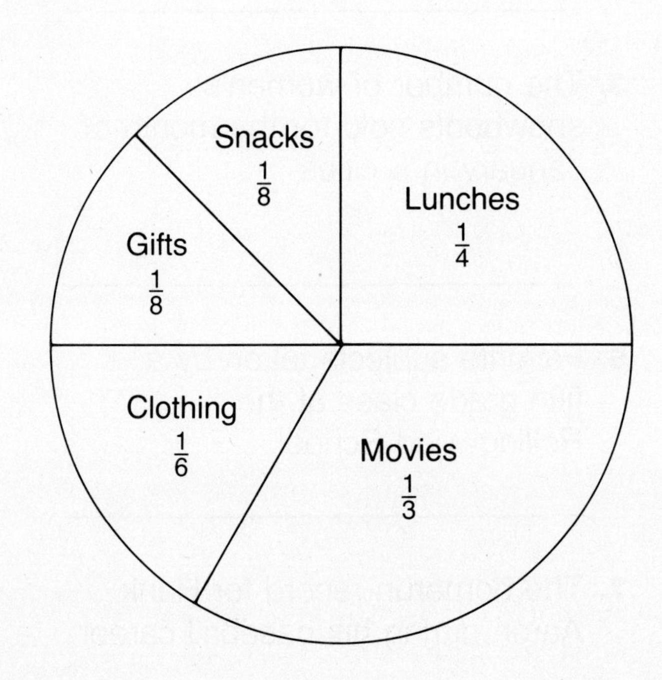

Ron spent $120. How much money was spent on each?

6. Lunches ______
7. Movies ______
8. Clothing ______
9. Gifts ______
10. Snacks ______
11. What did Ron spend more on, clothing or gifts? How much more?

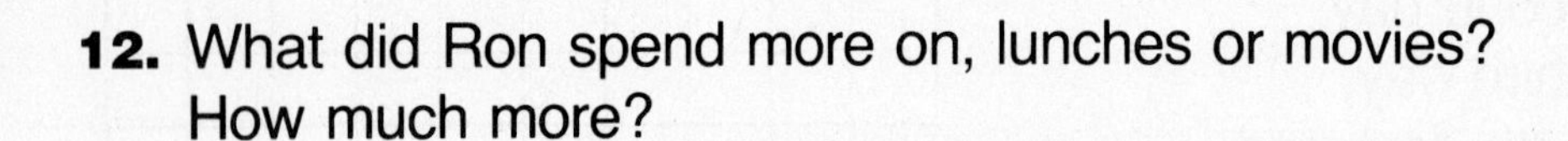

12. What did Ron spend more on, lunches or movies? How much more?

______

13. For every circle graph will the whole amount be $120? Explain.

______

14. What is the main idea of this circle graph?

______

NAME

SHARPEN YOUR SKILLS

Line Plots in Statistics

Use the line plot for math test scores to answer the questions.

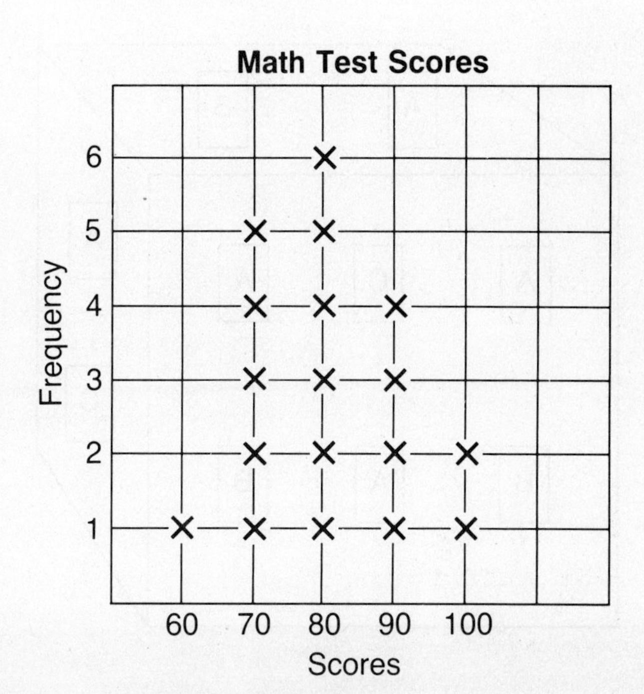

1. How many scores are less than 90?

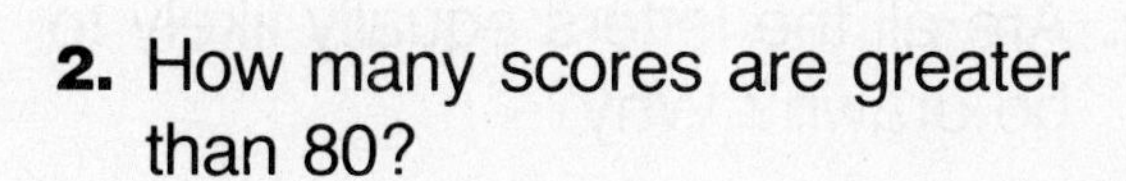

2. How many scores are greater than 80?

3. How many scores are there in all?

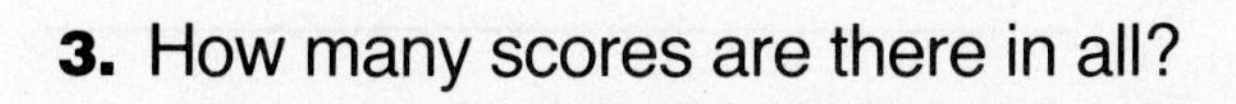

4. What is the mode?

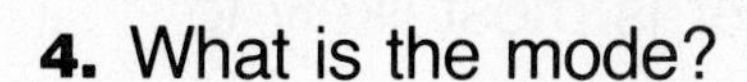

5. What is the median?

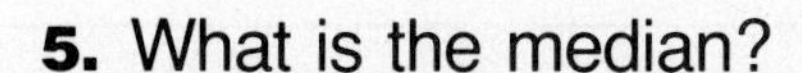

Use these data at the right to answer Exercises 6–8.

DATA: SCORES				
80	60	80	90	70
85	60	55	90	80

6. Fill in the frequency table below and then complete the line plot at the right.

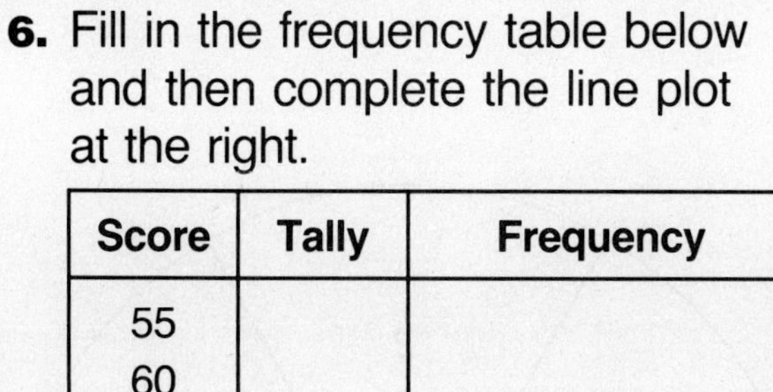

Score	Tally	Frequency
55		
60		
65		
70		
75		
80		
85		
90		

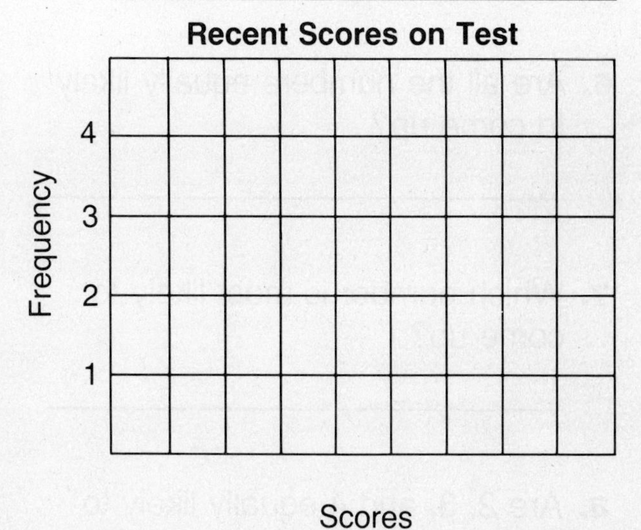

7. What is the mean?

8. What is the mode?

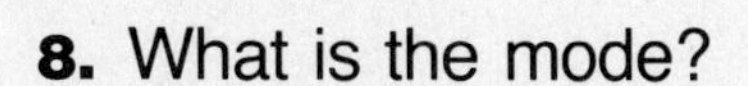

9. What is the median?

NAME

SHARPEN YOUR SKILLS

Outcomes

Letter cards are placed in a box.

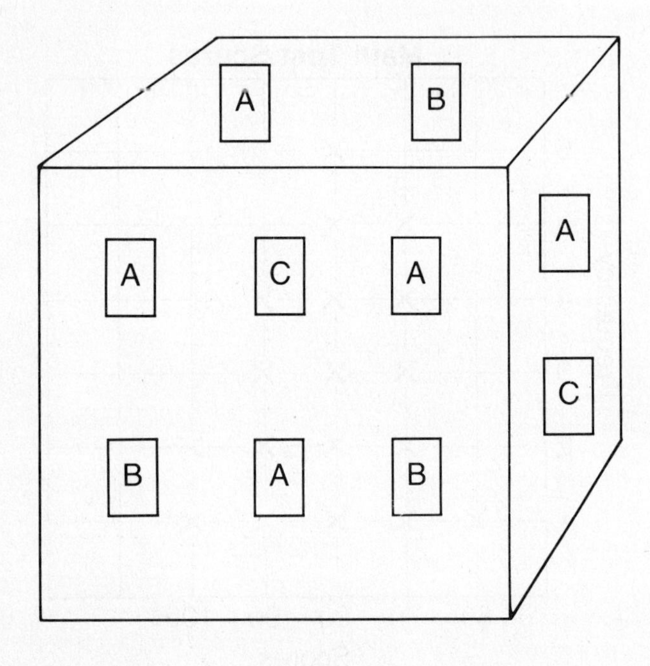

1. If one card is drawn at a time, list all the possible outcomes.

2. Are all the letters equally likely to be drawn? Why?

3. Which letter is most likely to be drawn?

Numbers are placed on a spinner.

4. Are B and C equally likely to be drawn? Why?

5. List the different outcomes.

6. Are all the numbers equally likely to come up?

7. Which number is most likely to come up?

8. Are 2, 3, and 4 equally likely to come up? Why?

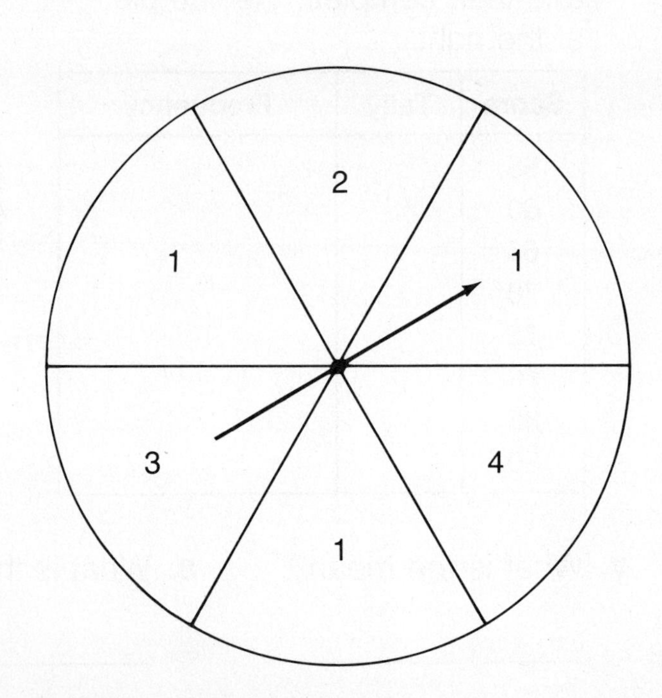

Experiments

Label pieces of paper as shown. Mix them and turn them over.

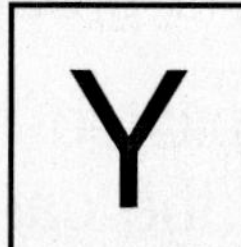

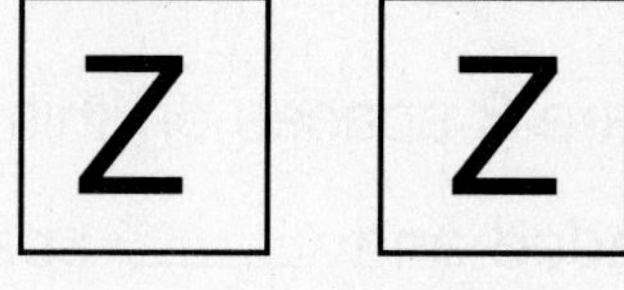

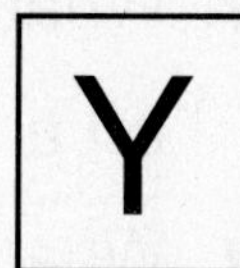

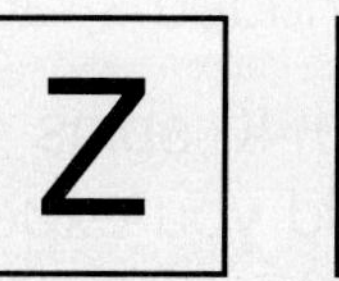

Write how many times you think you will choose each letter in 50 draws if you replace each letter and mix the pieces after each draw.

1. X ____________

2. Y ____________

3. Z ____________

Perform the experiment and record the results of 50 draws.

4. X ____________

5. Y ____________

6. Z ____________

7. Were your answers for Exercises 4–6 the same as in Exercises 1–3? Why might they be different?

____

____

Label 10 identical pieces of paper with the numbers 1–10.

8. In 40 draws, how many times do you think you will draw a number greater than 5?

____

9. Record the results of 40 draws. Was your prediction close to the actual result?

____

NAME ____________

SHARPEN YOUR SKILLS

Fractions and Probability

Arthur put 6 letter cards into a box.

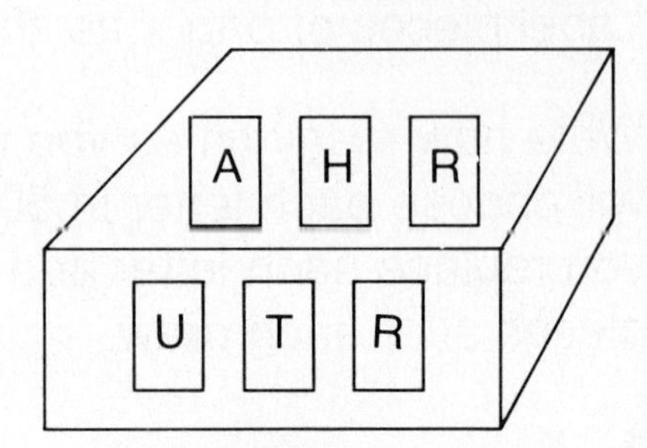

1. What is the probability of picking an R? $\frac{1}{3}$
2. If Arthur picked a card from the box 15 times, replacing the card each time, how many times is he likely to pick an R? ____

There are 8 spaces on this wheel. ____ spaces are shaded and ____ spaces are unshaded.

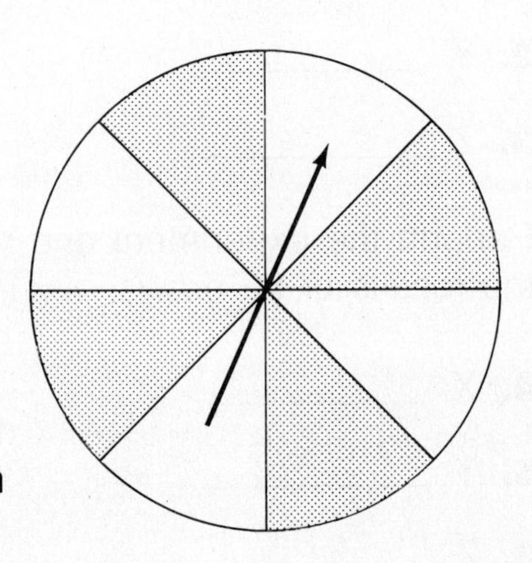

3. What is the probability of the spinner landing on a shaded space? ____
4. What is the probability of the spinner landing on an unshaded space? ____
5. If there are 40 spins of the wheel, how many times would you expect the spinner to land on a shaded space? ____

These letter tiles are placed in a box.

6. What is the probability of picking:

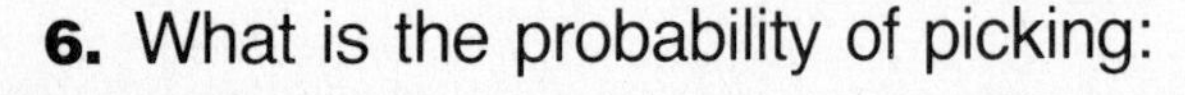

M ____ **S** ____ **I** ____ **P** ____

7. If 11 picks are made, replacing the tile each time, about how many would be I? ____

These shapes ○ □ ▭ △ ⌂ ⌓ appear on the 6 faces of a cube. When the cube is rolled, all the faces have the same chance of coming up.

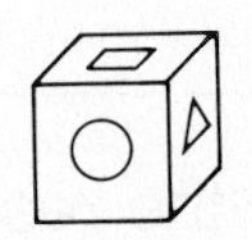

8. Find the probability of getting each result on one roll of the cube. ____________
9. What is the probability of getting a shape with 4 sides? 3 sides? 5 sides?

____

SHARPEN YOUR SKILLS

Solid Shapes

Use the objects shown below to complete the table.

Object	Faces	Vertices	Edges	Geometric Solid	Polyhedron (Yes/No)
A					
B					
C					
D					
E					
F					

A

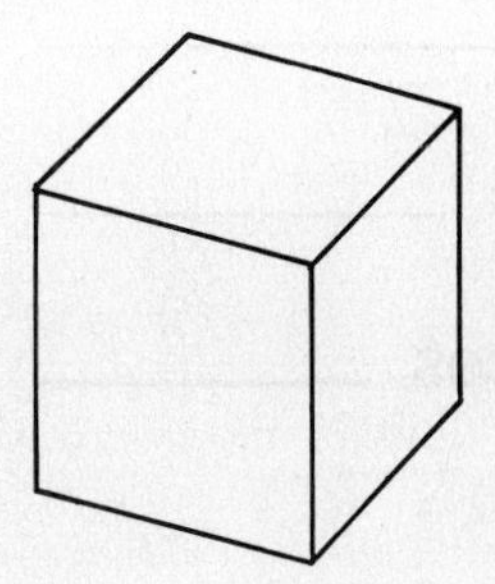

B

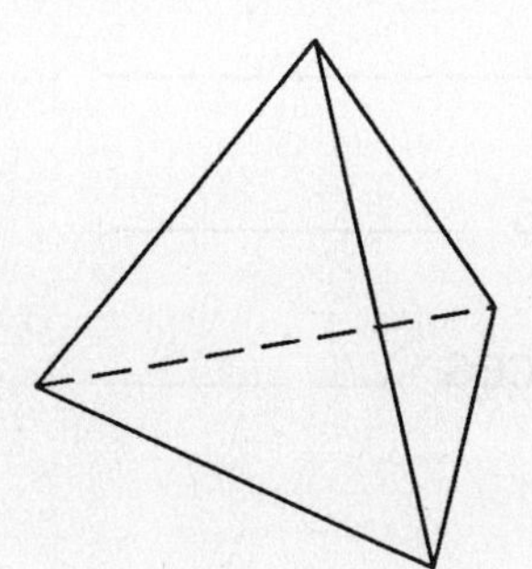

C

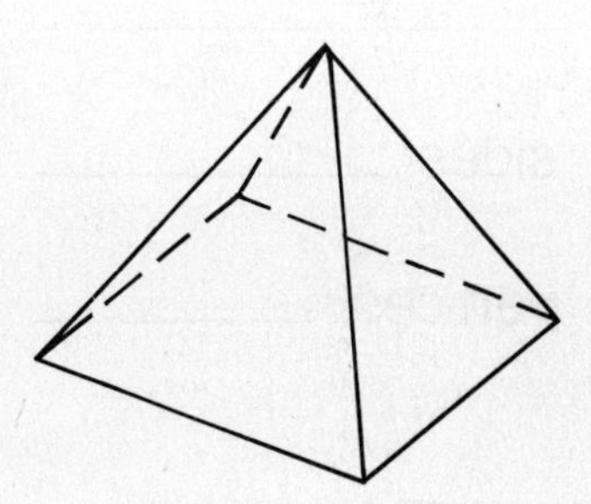

D

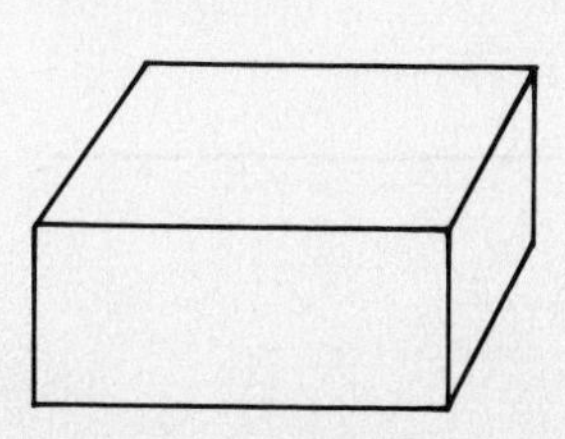

E

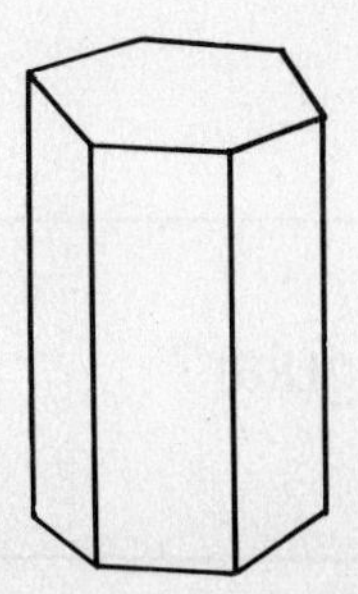

F

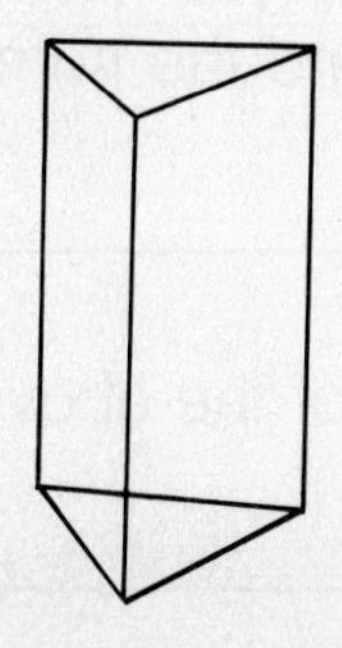

NAME

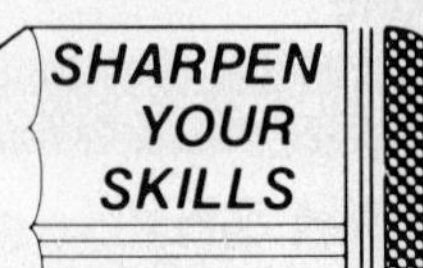

Polygons

Write which polygon you see.

Then tell how many sides
and vertices each polygon has.

1.

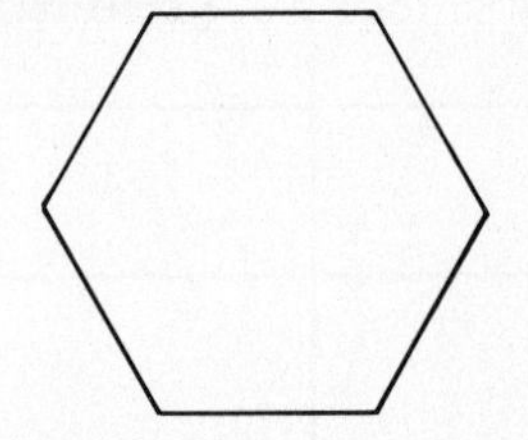

sides

vertices

2.

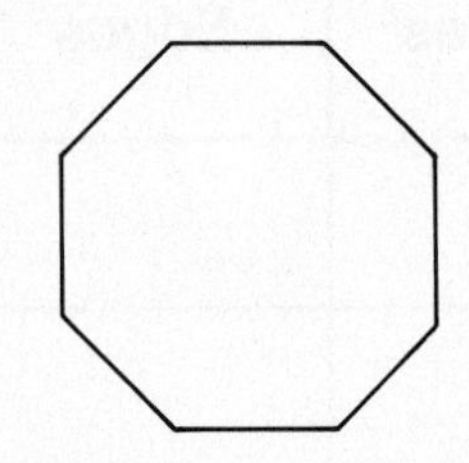

sides

vertices

3.

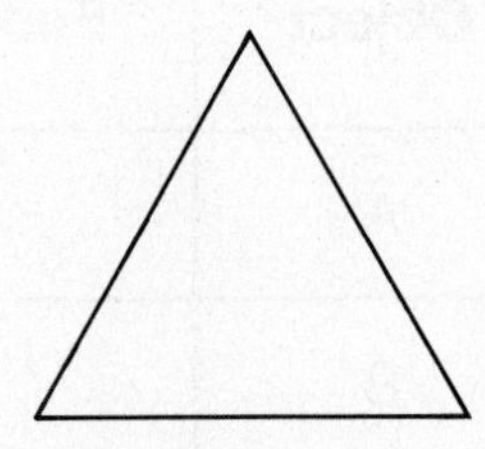

sides

vertices

4.

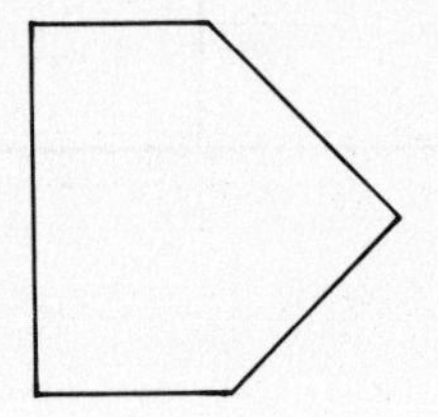

sides

vertices

5.

sides

vertices

6.

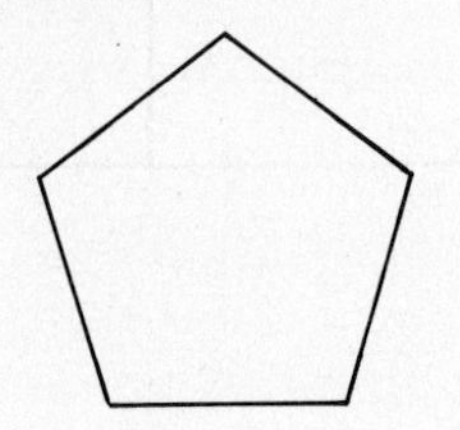

sides

vertices

Critical Thinking

7. What do you notice about the number of sides and the number of vertices?

8. Which of the above polygons are regular?

Basic Geometric Ideas

Name each line, segment, or ray.

1. L C

2. R P

3. G D

4. X M

5. B A

6. S W

Tell whether the lines are parallel lines or intersecting lines.

7.

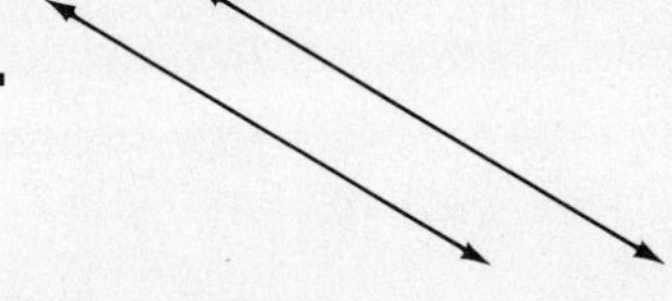

8.

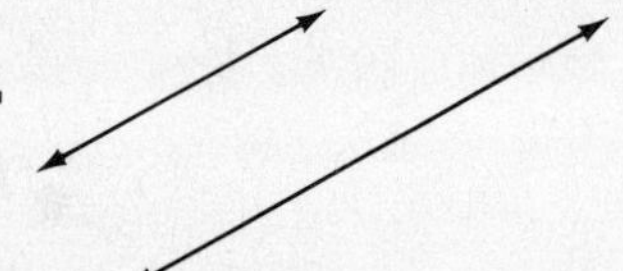

9.

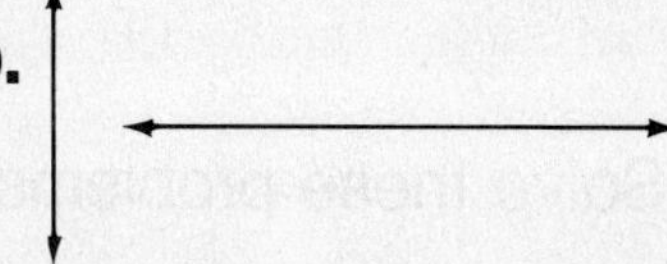

10.

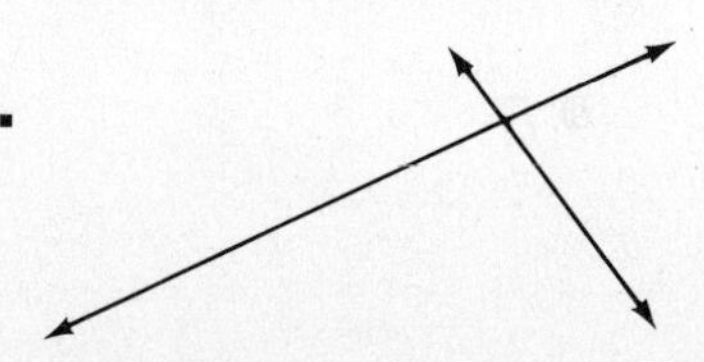

11.

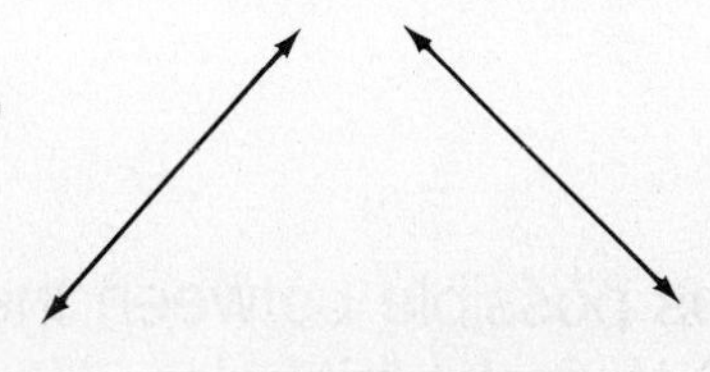

12.

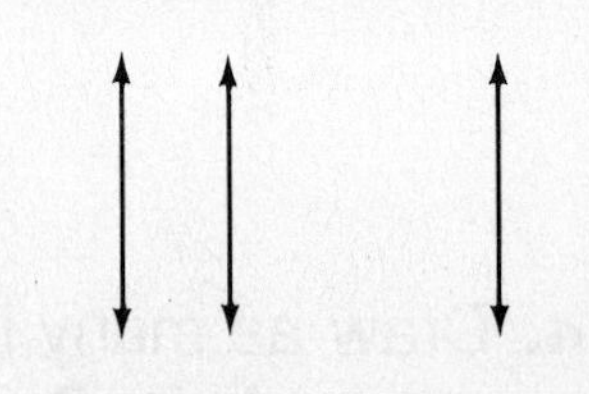

Name the geometric idea suggested by each picture.

13.

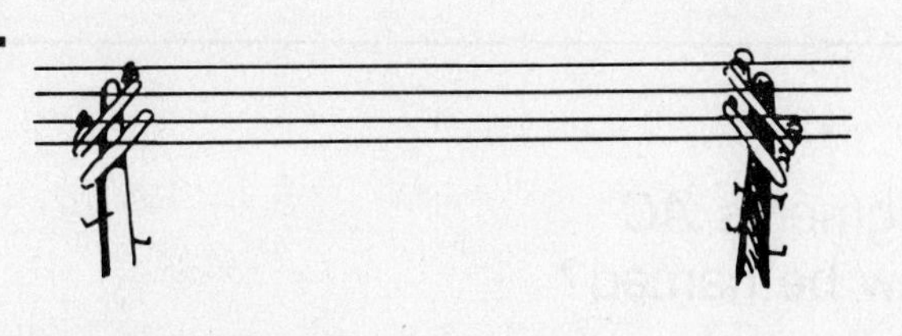

14.

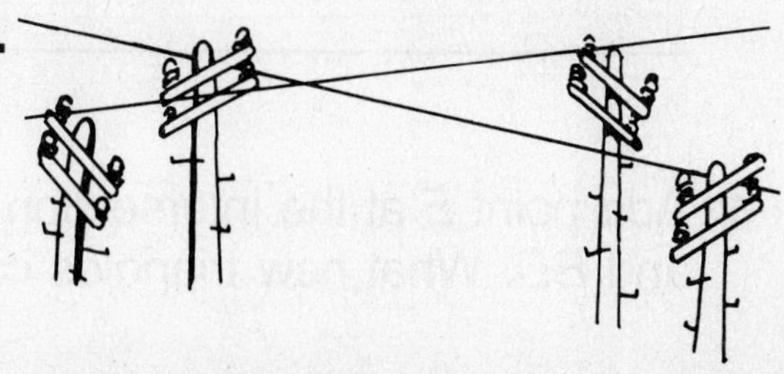

NAME

SHARPEN YOUR SKILLS

Triangles

Measure each side of each triangle. Then tell whether the triangle is equilateral, isosceles, or scalene.

1.

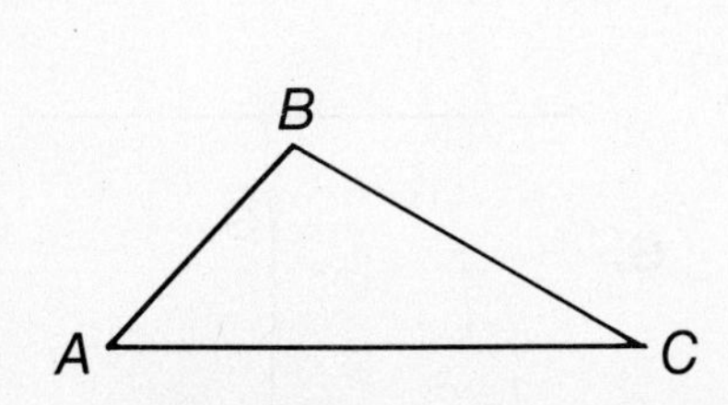

Segment *AB* ___ cm

Segment *AC* ___ cm

Segment *BC* ___ cm

____

2.

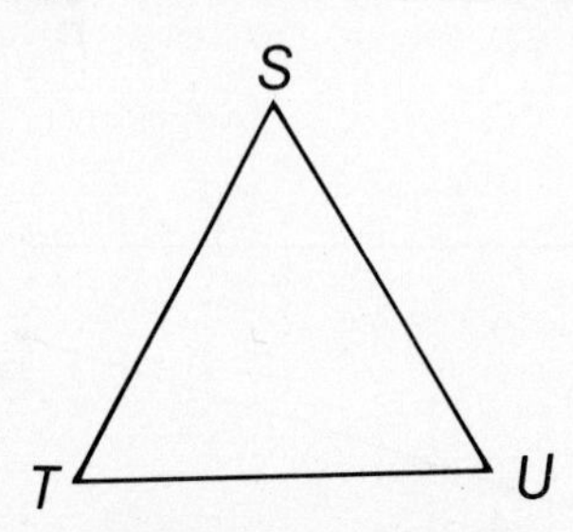

Segment *ST* ___ cm

Segment *SU* ___ cm

Segment *TU* ___ cm

____

3.

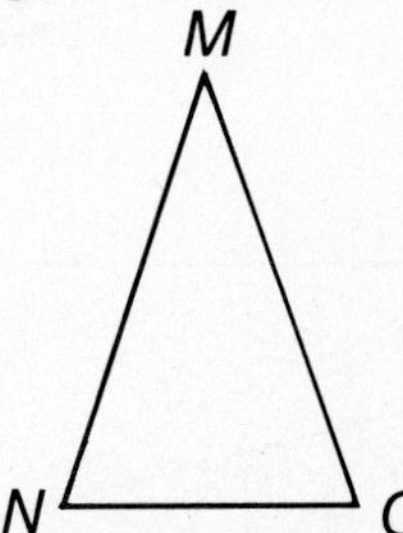

Segment *MN* ___ cm

Segment *MO* ___ cm

Segment *NO* ___ cm

____

Solve these problems.

B

A

C

D

4. Draw as many lines as possible between the points *A, B, C,* and *D* to make triangles. Name the triangles and tell whether they are *equilateral, isosceles,* or *scalene.*

____

____

5. Add point *E* at the intersection of segments *AC* and *BD*. What new triangles can now be named?

____

NAME

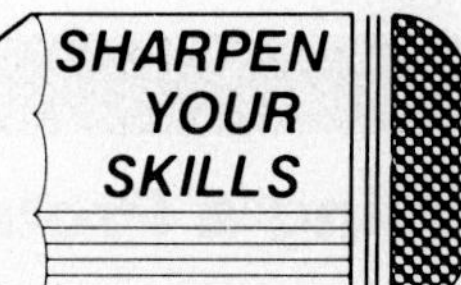

Angles

1. Complete the table.

Angle	Name	Vertex	Sides	Type
A, D, K				
P, M, B				
R, T, S				
T, X, Y				

For each problem, tell whether the angle formed by the clock hands is *right, acute,* or *obtuse.*

2.

____

3.

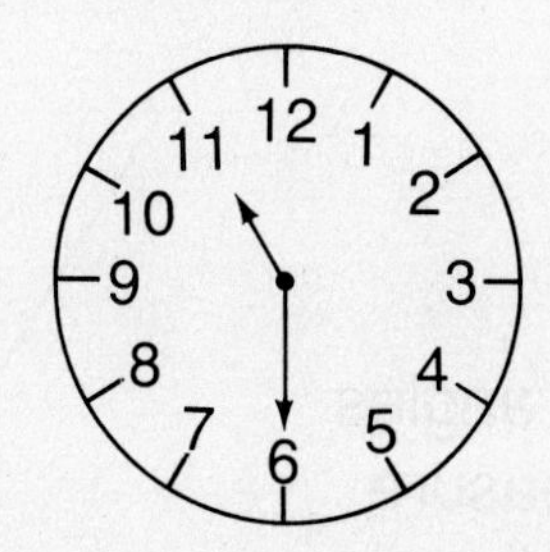

____

4.

____

NAME

SHARPEN YOUR SKILLS

Using a Protractor

Estimate the measure of each angle. Then use a protractor to find the measure. Tell whether the angle is *right, acute,* or *obtuse.*

1. ______________________ **2.** ______________________

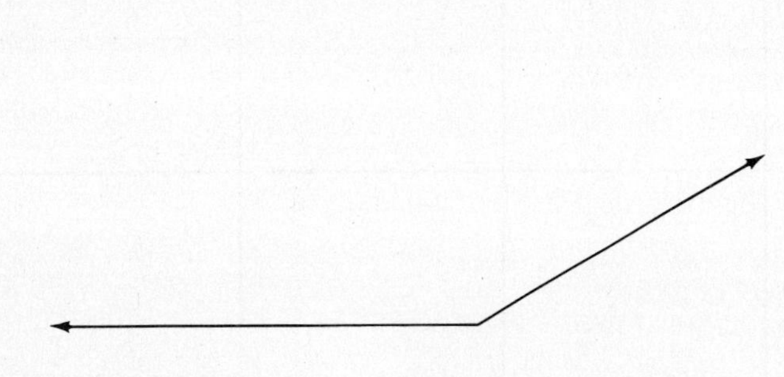

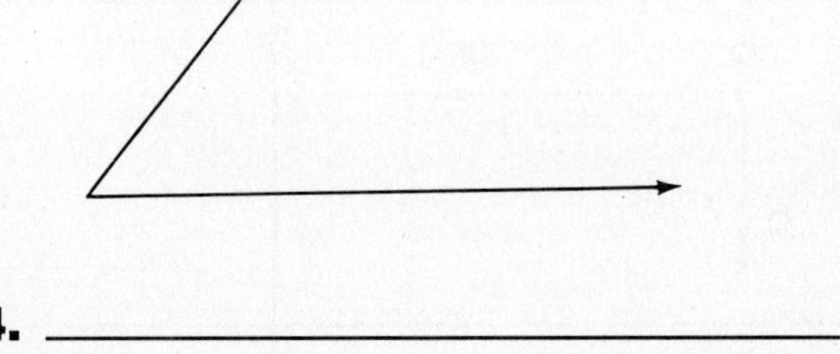

3. ______________________ **4.** ______________________

Use a protractor to draw an angle with a measure of

5. 160°

6. 35°

Solve the problem.

7. What is the sum of all the angles in the letters X and Y? Measure each angle in X, then find the sum. Repeat for Y.

____

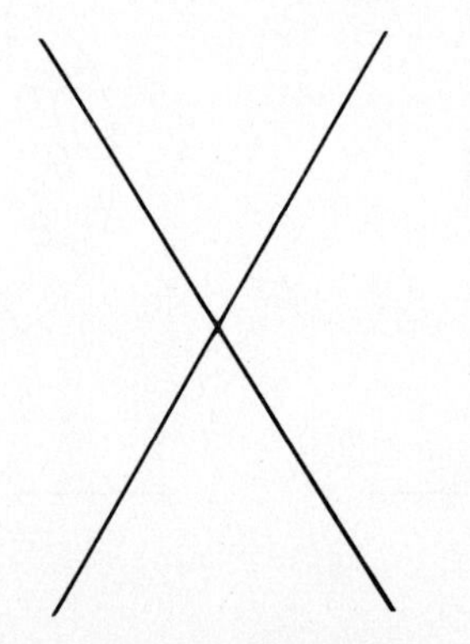

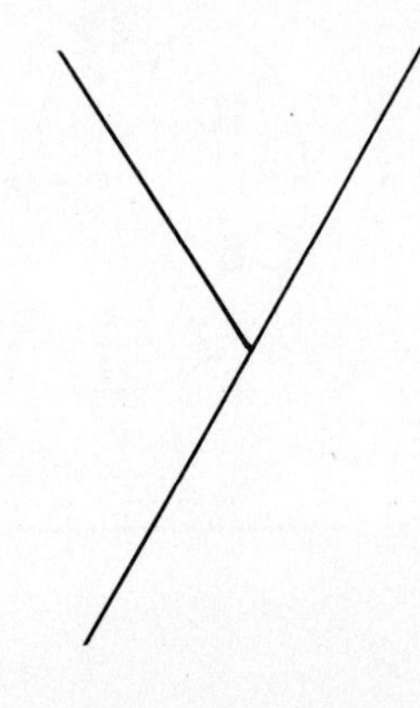

Quadrilaterals

Name each quadrilateral.

rhombus trapezoid square rectangle parallelogram

1.

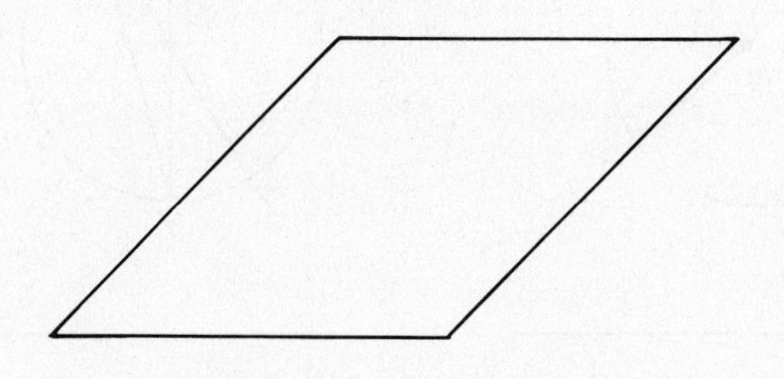

____

2.

____

3.

____

4.

____

5.

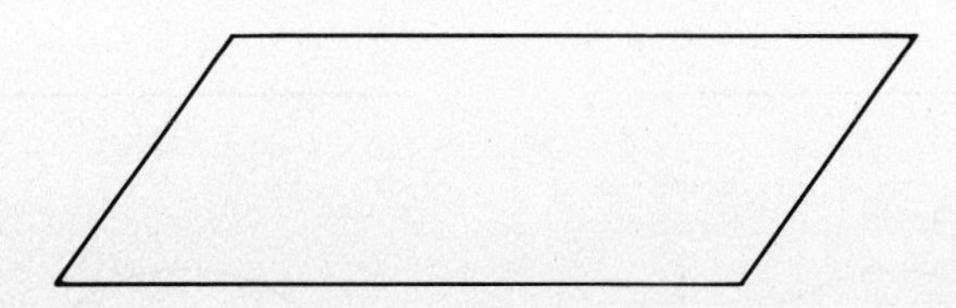

____

6.

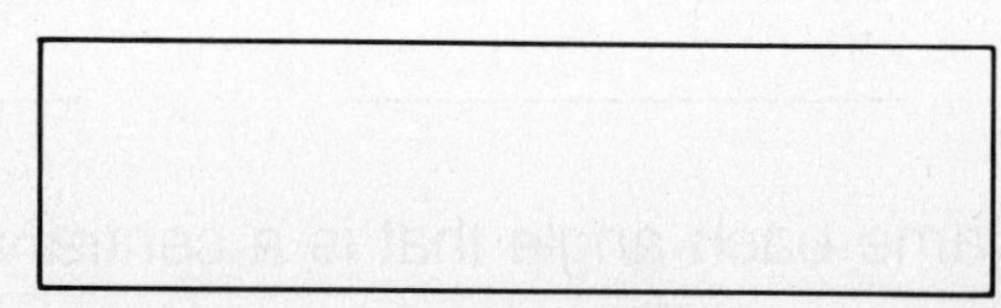

____

7.

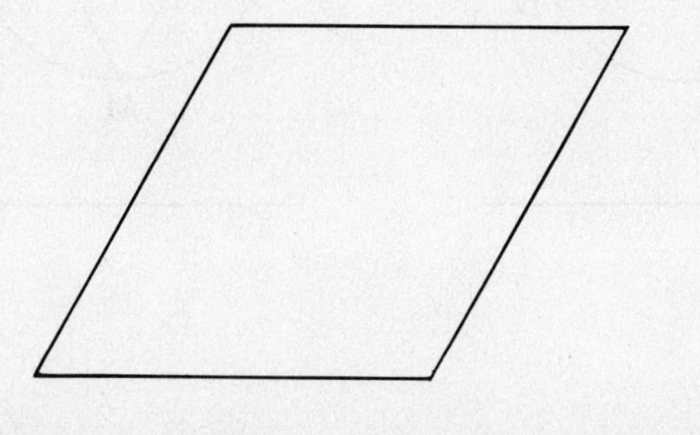

____

8.

____

SHARPEN YOUR SKILLS

Circles

Name each segment that is a radius.

1

2.

3.

Name each segment that is

4.

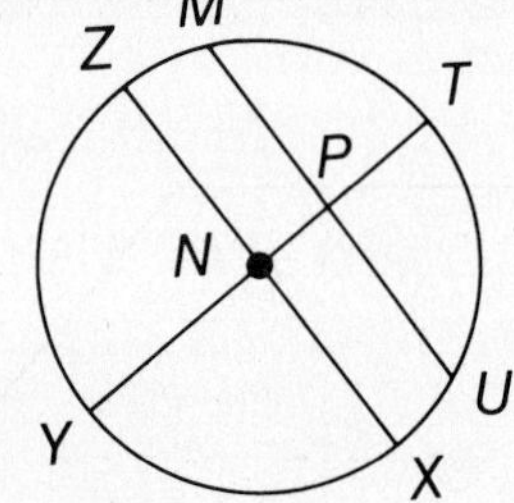

5.

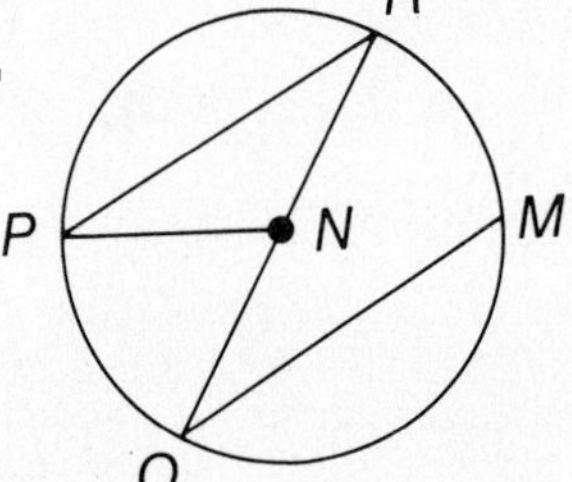

6.

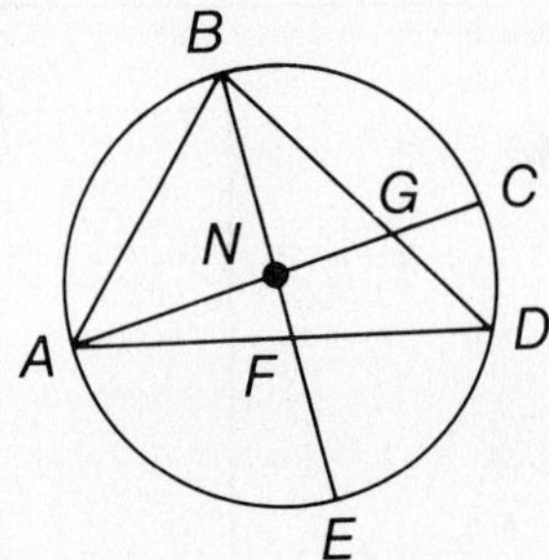

a diameter.

a chord.

Name each angle that is a central angle.

7.

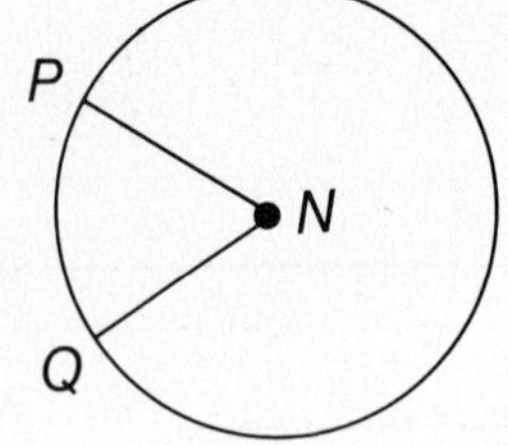

8.

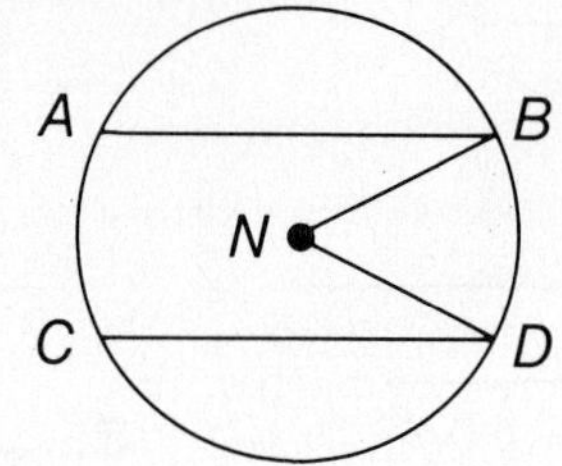

9.

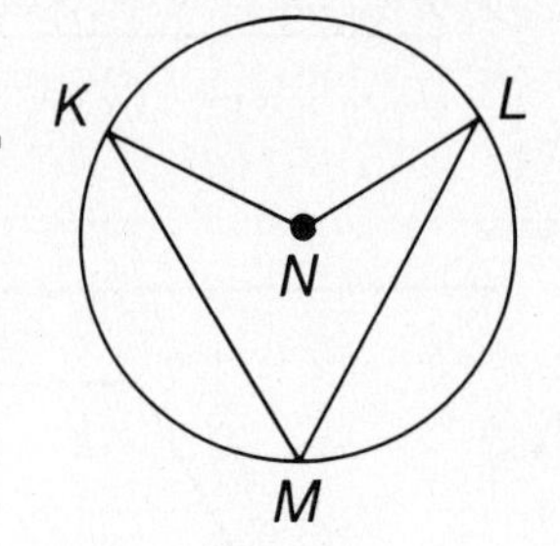

Choose diameter or radius to describe the problem.

10. The length of a spoke of a bicycle wheel. ______

NAME

SHARPEN YOUR SKILLS

Circumference

Estimation Estimate the circumference of each object.
Remember to label each unit of measure.

1. Complete the chart.

Object	Diameter	Circumference
Record	30 cm	
Penny	19 mm	
Birthday candle	6 mm	
Button	16 mm	
Kitchen table	110 cm	
Ring	18 mm	

Solve each problem.

2. Estimate the circumference of a bicycle wheel that has a 66 centimeter diameter.

____

3. The diameter of a circular watch is 2 centimeters. Estimate the circumference.

____

4. The distance from the center of a soup mug to the rim is 4 centimeters. What is the diameter of the mug?

____

5. The radius of a circular fish bowl is 14 centimeters. What is the diameter? Estimate the circumference.

____

Use after pages 474–475.

SHARPEN YOUR SKILLS

Congruent Figures

Tell whether the segments, angles, or figures are congruent. Write *yes* or *no.*

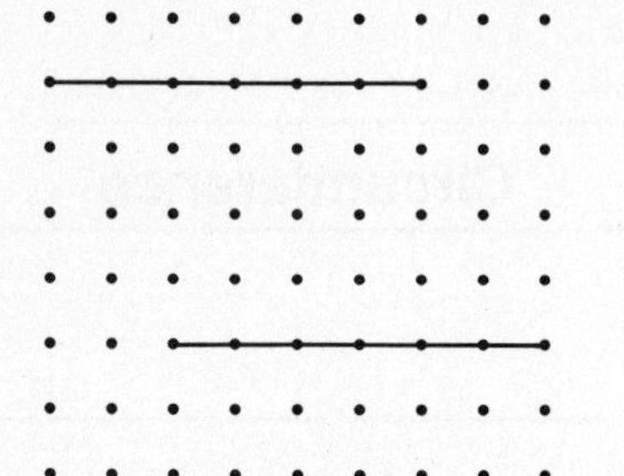

1. 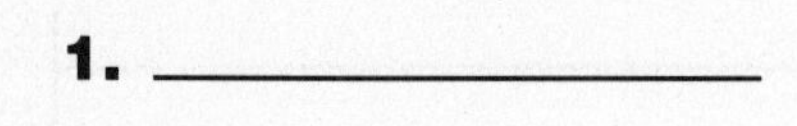

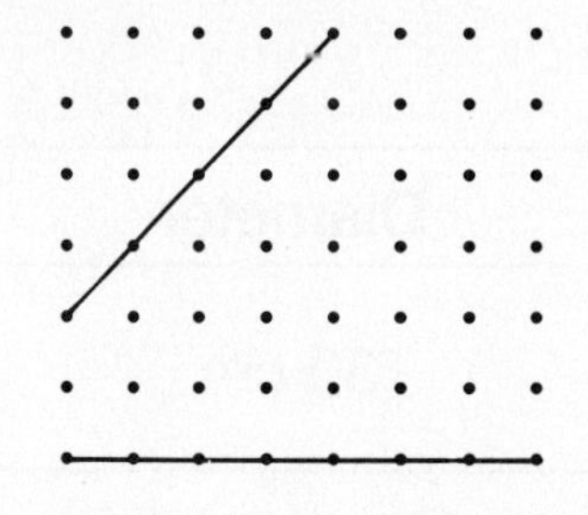

2. 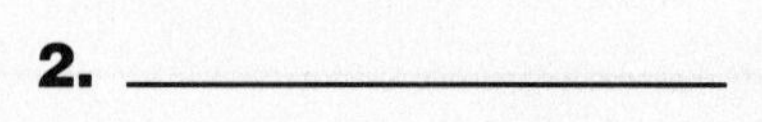

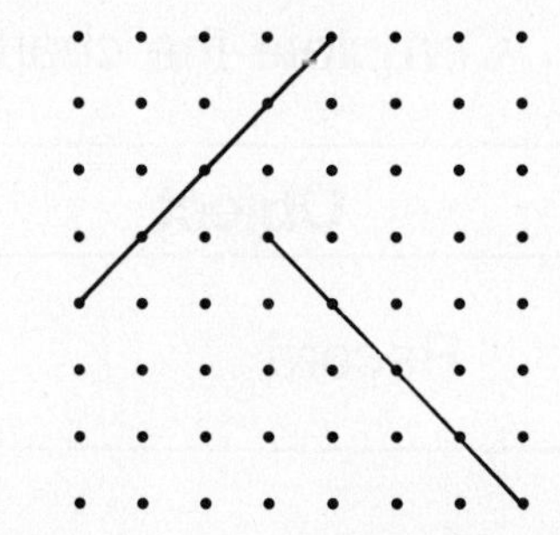

3. 

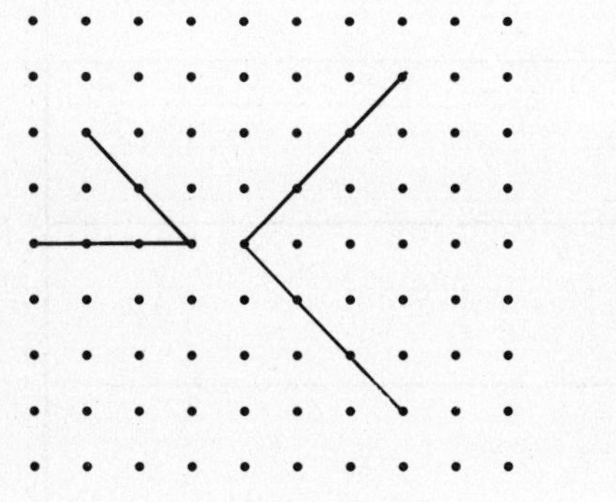

4. 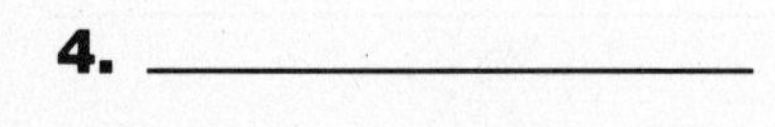

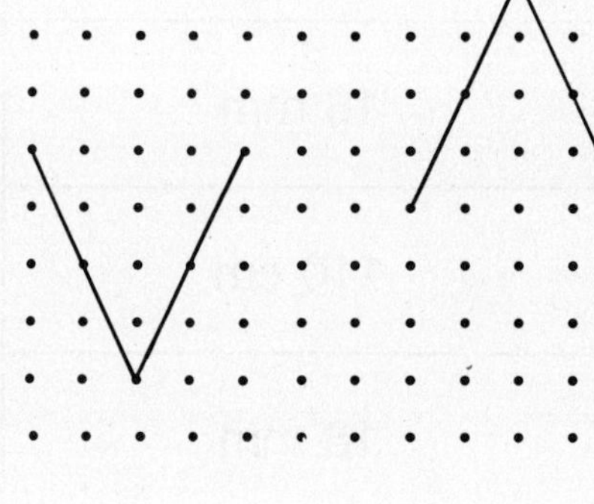

5. 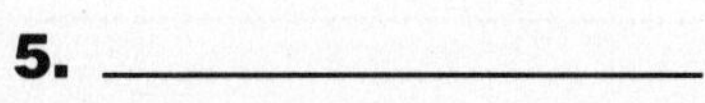

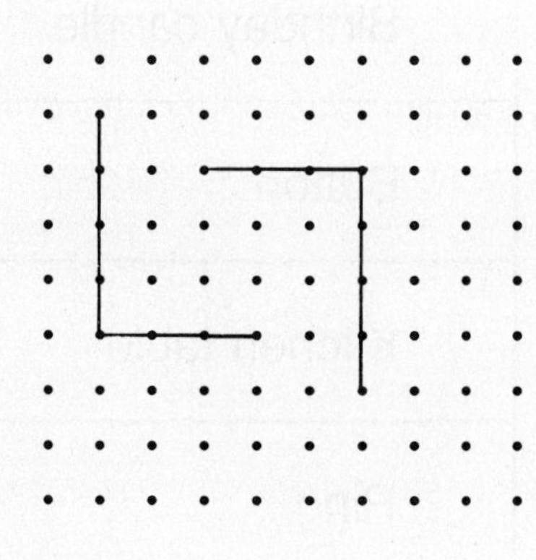

6. 

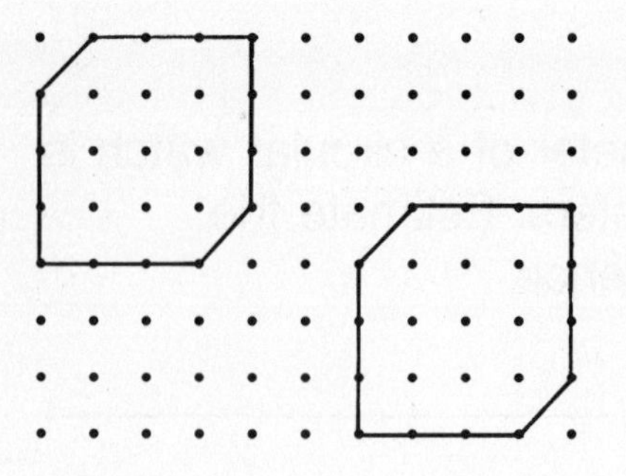

7. ______________________

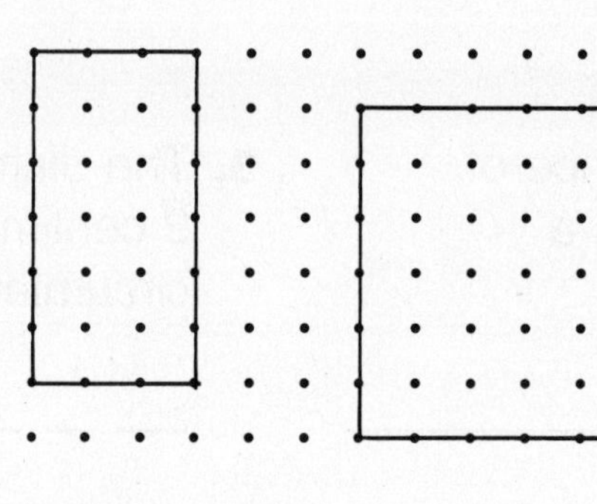

8. ______________________

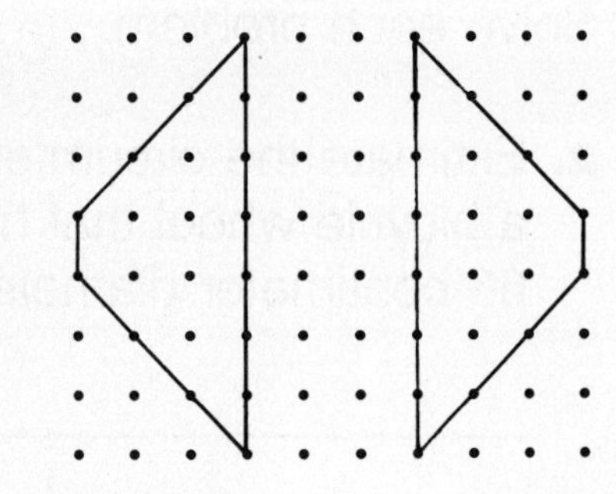

9. 

10. For each pair of congruent figures in Exercises 7–9, tell whether you would *flip, slide,* or *turn* one figure to put it exactly on top of the other figure.

____

NAME

SHARPEN YOUR SKILLS

Draw a Diagram

Trace the small figure. Then cut it out. Use the pattern to draw a picture to determine how many it takes to fill the large rectangular space.

1.

____

2.

____

3.

____

Use after pages 478–479.

NAME

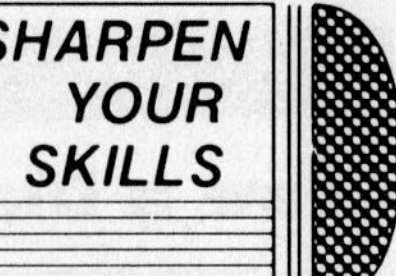

Symmetry

Draw all the lines of symmetry. How many lines of symmetry does each figure have?

1. 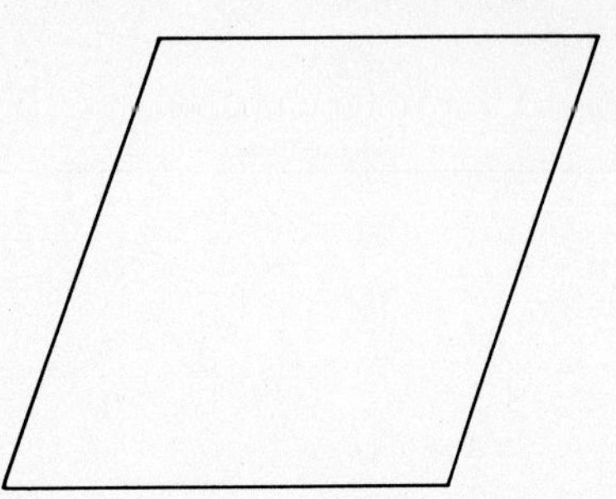 ____

2. 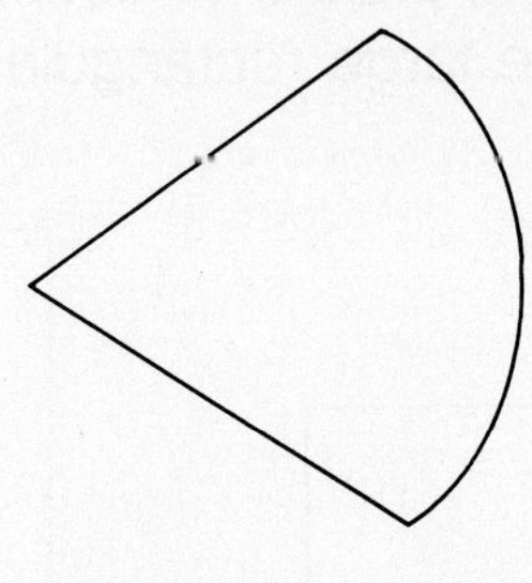 ____

3. 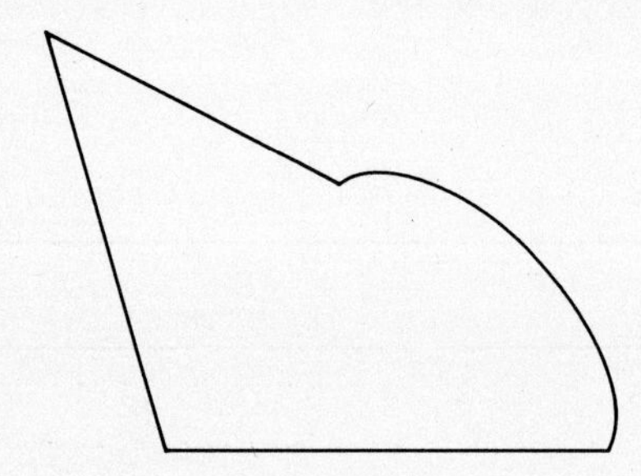 ____

4.  ____

5.  ____

6.  ____

The dotted line is the line of symmetry. Draw the other part of the figure.

7.

8. 

9. 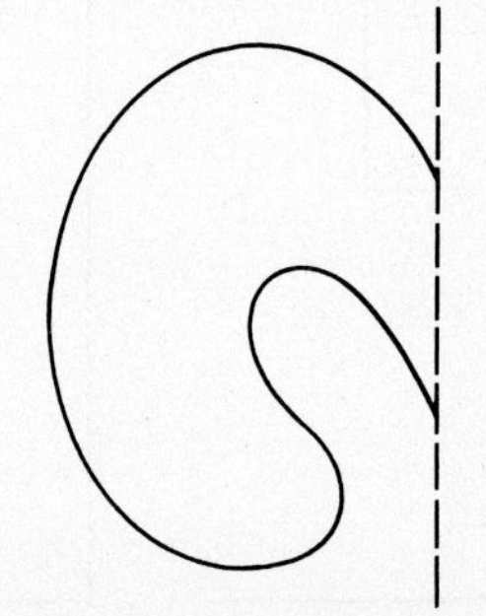

NAME

SHARPEN YOUR SKILLS

Similar Figures

Tell whether the figures are similar.
Write *yes* or *no.*

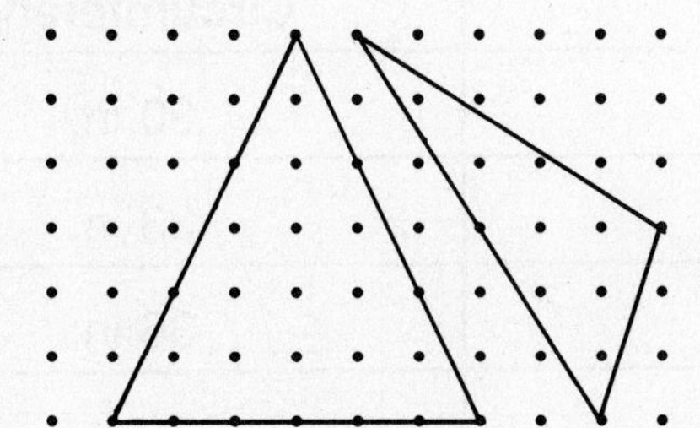

1. ______

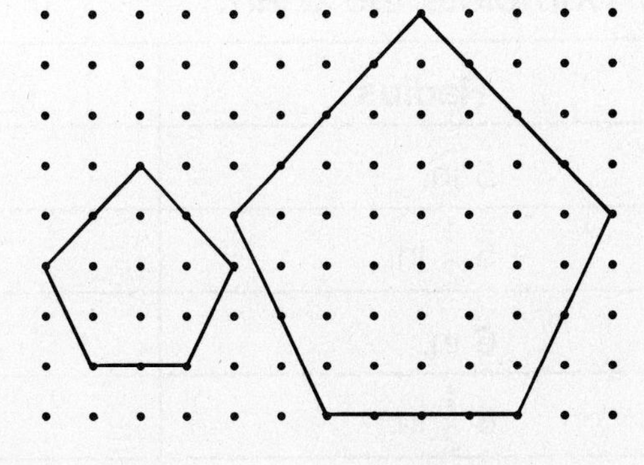

2. ______

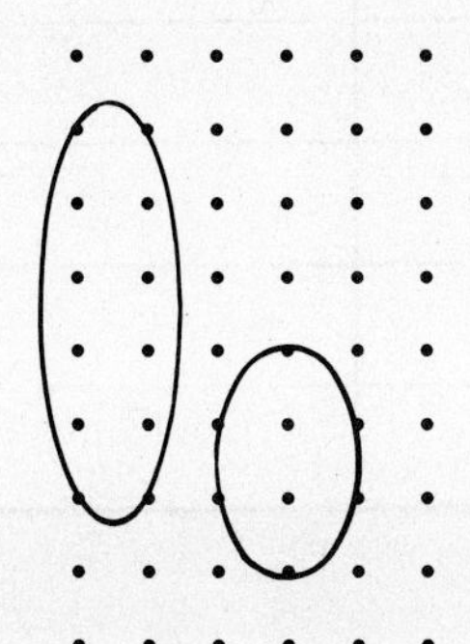

3. ______

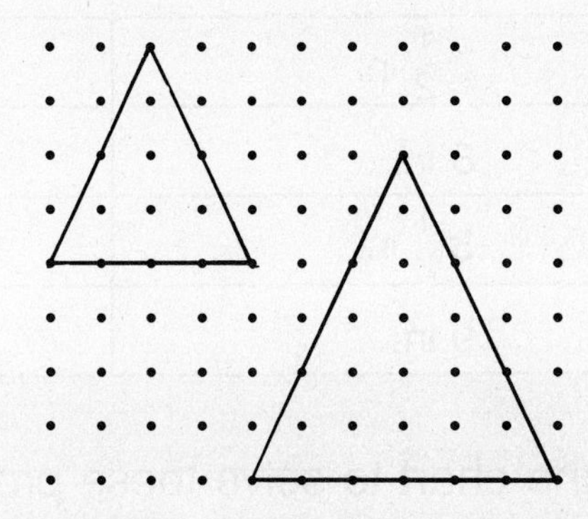

4. ______

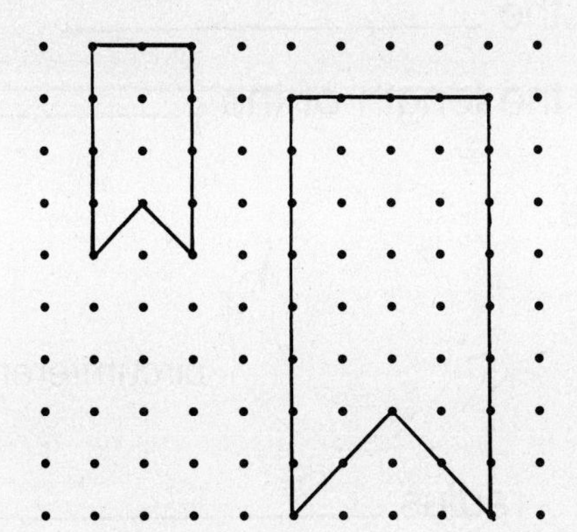

7. ______

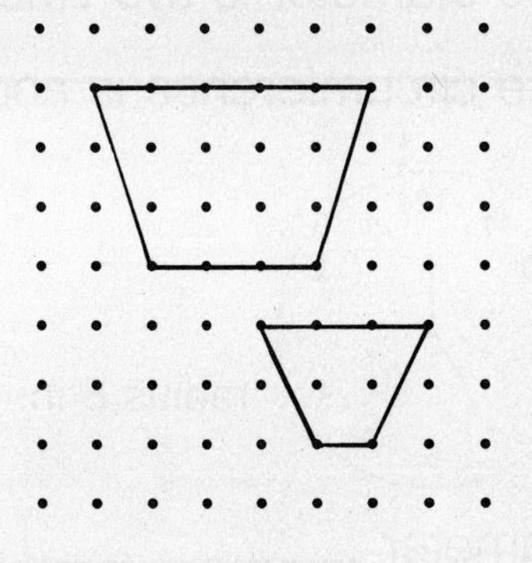

6. ______

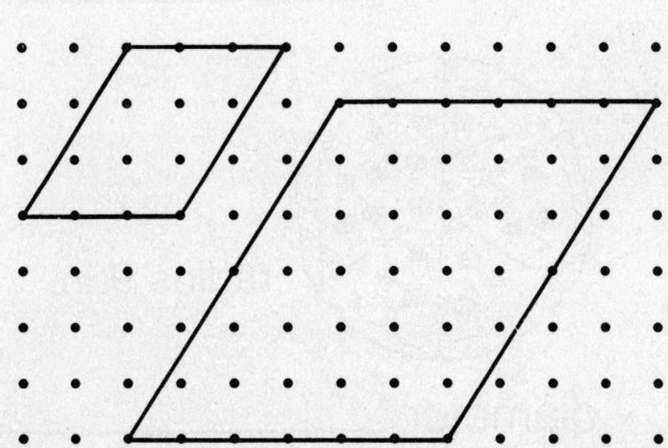

5. ______

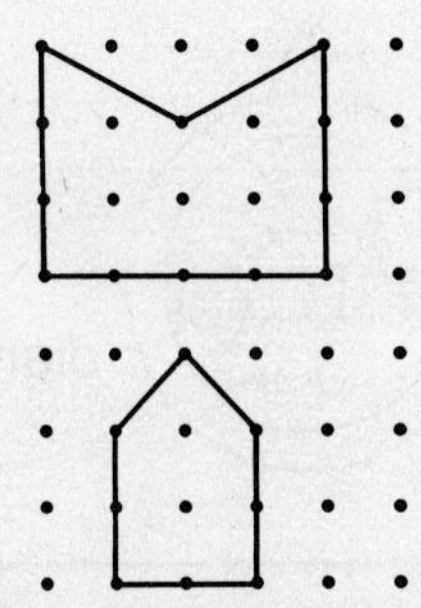

8. ______

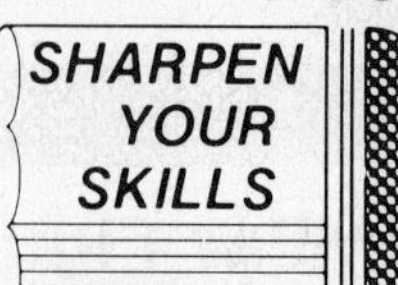

Find a Pattern

1. Complete the chart.

Radius	Diameter	Circumference
5 in.	10 in.	30 in.
$5\frac{1}{2}$ in.	11 in.	33 in.
6 in.	12 in.	36 in.
$6\frac{1}{2}$ in.		
7 in.		
$7\frac{1}{2}$ in.		
8 in.		
$8\frac{1}{2}$ in.		
9 in.		

Use the chart to solve these problems.

2. The diameter is two times the length of the ________________ .

3. The circumference is about three times the length of the ________________ .

4. 

radius 8 in.

diameter ________________

circumference ________________

5. 

circumference 30 in.

radius ________________

diameter ________________

6. 

diameter 14 in.

radius ________________

circumference ________________

7. 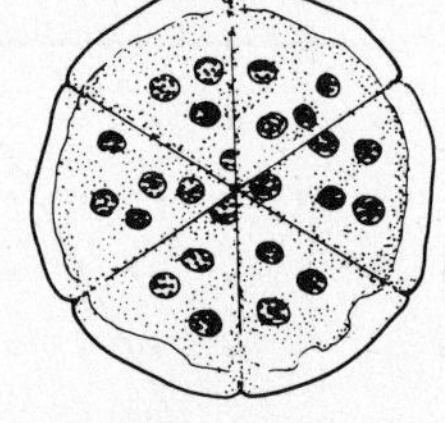

radius 9 in.

diameter ________________

circumference ________________

Use after pages 484–485.

NAME

SHARPEN YOUR SKILLS

Meaning of Ratio

Write a ratio in two ways for each exercise.

1.

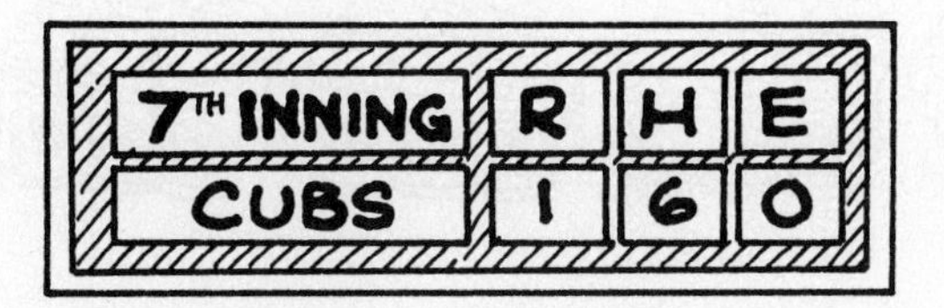

Hits to Runs

2.

Markers to Cents

3.

Tomatoes to Kilograms

4.

Roses to Cents

For each exercise, write a ratio and label the numbers.

5. Yesterday, 3 out of 29 students were absent.

6. It takes 5 nickels to equal 1 quarter.

7. A hummingbird beats its wings 80 times in 1 second.

8. Some people sleep 55 hours out of every 168 hours.

SHARPEN YOUR SKILLS

Equal Ratios

Use the pictures to write a list of equal ratios for each exercise.

1. There are 6 cans of juice in each carrier.

Cans to Carriers 6:1 12:___ ___:3 24:___ ___:5

2. Each game set has 2 tennis rackets and 3 tennis balls.

Tennis rackets to Tennis balls 2:3 4:___ ___:9 8:___ ___:___

3. Each flag of Chicago has 4 stars and 2 stripes.

Stars to Stripes 4:2 ___:4 12:___ ___:___ 20:___

4. Each sofa has 5 back cushions and 2 seat cushions.

Back cushions to Seat cushions 5:2 ___:4 15:___ ___:8 25:___

Use after pages 498–499.

SHARPEN YOUR SKILLS

Finding Equal Ratios

Complete each list of equal ratios.

1. $\frac{6}{4} = \frac{\square}{8} = \frac{18}{\square} = \frac{\square}{16} = \frac{30}{\square}$

2. $\frac{9}{15} = \frac{18}{\square} = \frac{\square}{45} = \frac{\square}{60} = \frac{45}{\square}$

Write 3 more ratios equal to the ratio given.

3. $\frac{2}{5} = \frac{\underline{\quad}}{20} = \frac{10}{\underline{\quad}} = \frac{\underline{\quad}}{30}$

4. $\frac{5}{7} = \frac{10}{\underline{\quad}} = \frac{\underline{\quad}}{28} = \frac{25}{\underline{\quad}}$

5. $\frac{4}{3} = \frac{12}{\underline{\quad}} = \frac{\underline{\quad}}{12} = \frac{\underline{\quad}}{18}$

6. $\frac{4}{9} = \frac{\underline{\quad}}{18} = \frac{\underline{\quad}}{27} = \frac{16}{\underline{\quad}}$

7. $\frac{8}{4} = \frac{16}{\underline{\quad}} = \frac{\underline{\quad}}{12} = \frac{\underline{\quad}}{16}$

8. $\frac{3}{4} = \frac{\underline{\quad}}{20} = \frac{21}{\underline{\quad}} = \frac{\underline{\quad}}{36}$

9. $\frac{8}{9} = \frac{24}{\underline{\quad}} = \frac{40}{\underline{\quad}} = \frac{\underline{\quad}}{54}$

10. $\frac{4}{5} = \frac{12}{\underline{\quad}} = \frac{\underline{\quad}}{20} = \frac{\underline{\quad}}{30}$

11. $\frac{9}{3} = \frac{\underline{\quad}}{6} = \frac{\underline{\quad}}{9} = \frac{\underline{\quad}}{15}$

12. $\frac{13}{5} = \frac{\underline{\quad}}{10} = \frac{\underline{\quad}}{20} = \frac{\underline{\quad}}{30}$

13. $\frac{12}{6} = \frac{24}{\underline{\quad}} = \frac{\underline{\quad}}{18} = \frac{\underline{\quad}}{30}$

14. $\frac{8}{12} = \frac{24}{\underline{\quad}} = \frac{40}{\underline{\quad}} = \frac{\underline{\quad}}{72}$

Solve each problem using equal ratios.

15. Kevin sleeps 9 hours each night. How many hours does he sleep in 7 nights?

____

____

16. The coach uses 3 footballs for every 5 players at practice. How many footballs are used for 20 players?

____

____

NAME

SHARPEN YOUR SKILLS

Proportions

Use cross-products to find which ratios form a proportion. Then shade the areas that have equal ratios. You will see the shape of a useful tool you use every day.

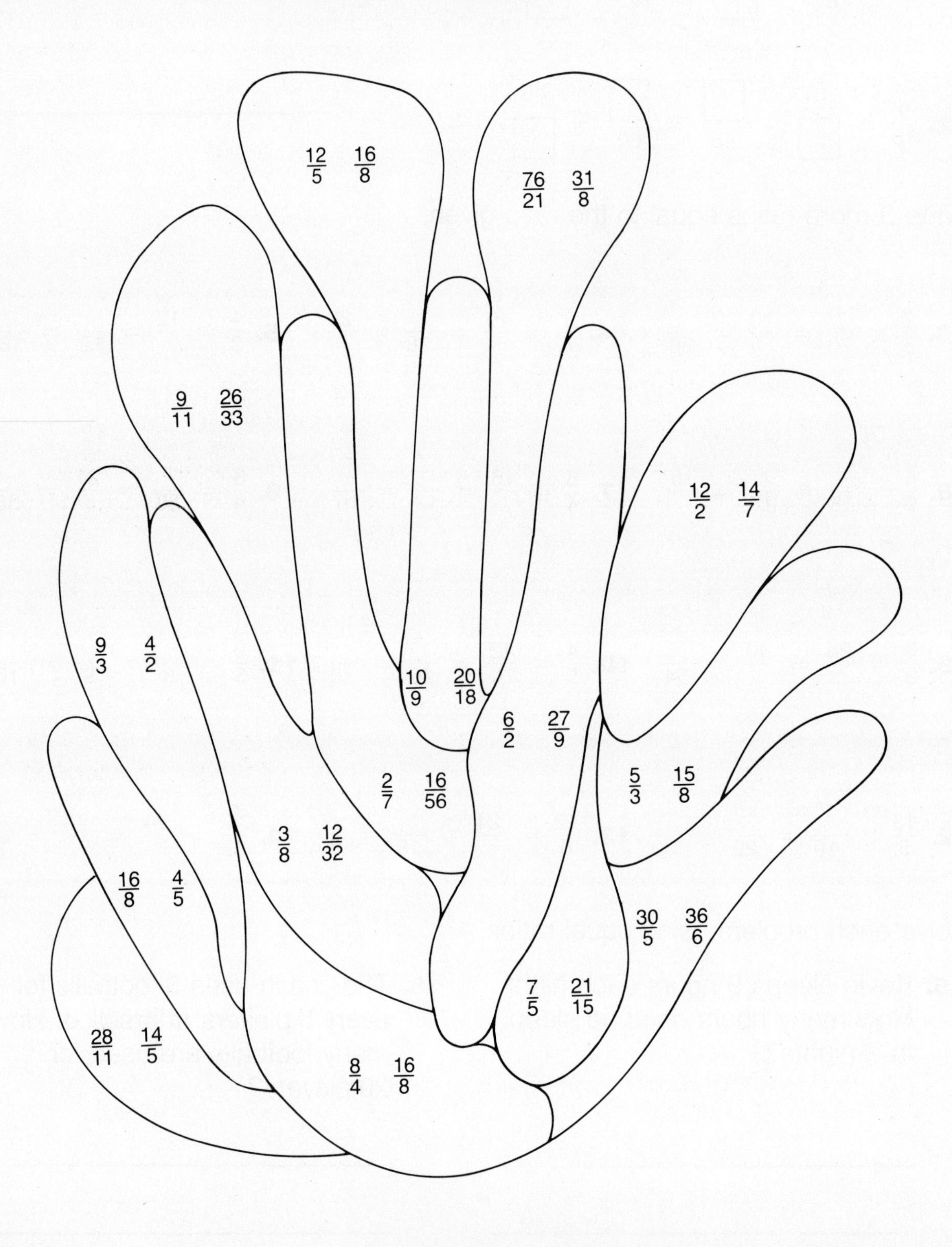

NAME

SHARPEN YOUR SKILLS

Write an Equation

Write a proportion to solve each problem.
Remember to use the same order in both ratios.

1. A map has a scale in which 1 inch represents 6 miles. What is the distance in miles of 7 inches?

____

2. Lorraine is making a scale drawing of her house. The house is 60 feet long and 36 feet wide. If she uses 12 inches to represent the width, how many inches should she use to represent the length?

____

3. Roberto bought $6.05 worth of tomato sauce. The cost per can is $0.55. How many cans did he buy?

____

4. A car travels 360 miles on 12 gallons of gas. How many miles will it travel on 1 gallon of gas?

____

5. An airplane can travel 450 miles in 1 hour. How far will that plane travel in 6 hours?

____

6. If 2 trucks hold 8 people, how many trucks would be needed to carry 256 people?

____

7. Renada reads about 150 pages each week. About how many pages does she read in 1 year?

____

8. Jim types about 6 pages an hour. How long will it take him to type 45 pages?

____

Use after pages 504–505.

SHARPEN YOUR SKILLS

Ratios and Percents

Write each ratio as a percent.

1. $\frac{34}{100}$ ______ **2.** $\frac{40}{100}$ ______ **3.** $\frac{10}{100}$ ______ **4.** $\frac{99}{100}$ ______ **5.** $\frac{67}{100}$ ______

6. 6 to 100 ______ **7.** 55 to 100 ______ **8.** 29 to 100 ______ **9.** 3 to 100 ______ **10.** 12 to 100 ______

11. 33:100 ______ **12.** 1:100 ______ **13.** 66:100 ______ **14.** 75:100 ______ **15.** 25:100 ______

Write a ratio for each percent.

16. 33% ______ **17.** 5% ______ **18.** 45% ______ **19.** 89% ______ **20.** 10% ______

21. 20% ______ **22.** 15% ______ **23.** 75% ______ **24.** 98% ______ **25.** 56% ______

Solve each problem.

26. Fred found that 56 out of 100 students liked baseball more than football. What percent of them liked baseball better?

______

27. What percent of the students liked football better than baseball?

______

SHARPEN YOUR SKILLS

Percents and Fractions

Write each percent as a fraction in lowest terms.

1. 34% ______ **2.** 55% ______ **3.** 25% ______ **4.** 60% ______ **5.** 40% ______

6. 65% ______ **7.** 20% ______ **8.** 90% ______ **9.** 75% ______ **10.** 12% ______

Write each fraction as a percent.

11. $\frac{43}{100}$ ______ **12.** $\frac{7}{100}$ ______ **13.** $\frac{4}{5}$ ______ **14.** $\frac{7}{10}$ ______ **15.** $\frac{56}{100}$ ______

16. $\frac{3}{50}$ ______ **17.** $\frac{1}{10}$ ______ **18.** $\frac{2}{5}$ ______ **19.** $\frac{9}{20}$ ______ **20.** $\frac{21}{50}$ ______

Use the circle graph to answer each question.

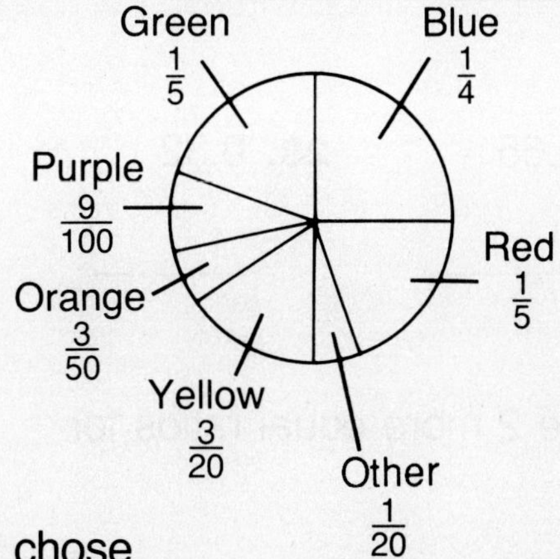

What percent of the students chose

21. purple? ______ **22.** yellow? ______ **23.** red? ______

24. blue? ______ **25.** green? ______ **26.** orange? ______

SHARPEN YOUR SKILLS

Percents and Decimals

Write each percent as a decimal.

1. 53% ______ **2.** 46% ______ **3.** 97% ______ **4.** 52% ______ **5.** 50% ______

6. 8% ______ **7.** 75% ______ **8.** 10% ______ **9.** 12% ______ **10.** 60% ______

11. 67% ______ **12.** 33% ______ **13.** 18% ______ **14.** 100% ______ **15.** 3% ______

Write each decimal as a percent.

16. 0.56 ______ **17.** 0.27 ______ **18.** 0.2 ______ **19.** 0.05 ______ **20.** 0.39 ______

21. 0.09 ______ **22.** 0.65 ______ **23.** 0.32 ______ **24.** 0.98 ______ **25.** 0.76 ______

Mixed Practice Write 2 more equal ratios for each ratio below.

26. $\frac{3}{5} = \frac{\quad}{15} = \frac{12}{\quad}$

27. $\frac{5}{8} = \frac{15}{\quad} = \frac{25}{\quad}$

28. $\frac{7}{16} = \frac{21}{\quad} = \frac{\quad}{80}$

29. $\frac{4}{9} = \frac{20}{\quad} = \frac{32}{\quad}$

30. $\frac{6}{7} = \frac{42}{\quad} = \frac{\quad}{91}$

31. $\frac{9}{11} = \frac{\quad}{88} = \frac{\quad}{264}$

NAME

SHARPEN YOUR SKILLS

Finding a Percent of a Number

Find the percent of each number.

1. 50% of 30 ______

2. 30% of 90 ______

3. 80% of 65 ______

4. 5% of 120 ______

5. 2% of 50 ______

6. 25% of 160 ______

7. 8% of 75 ______

8. 45% of 200 ______

9. 18% of 400 ______

10. 33% of 200 ______

11. 90% of 170 ______

12. 100% of 987 ______

Number Sense Write <, >, or = in each ◯.

13. 90% of 50 ◯ 50

14. 100% of 70 ◯ 70

15. 110% of 52 ◯ 52

16. 100% of 32 ◯ 34

17. 50% of 60 ◯ 28

18. 25% of 16 ◯ 5

Solve each problem.

19. Apples at the Fresh-Buy Farm Stand are normally $5 per bag. How much less will they be if they are reduced by 15%?

______

20. What will the cost of a bag of apples be after the 15% is taken off?

______

21. Peaches at the farm stand normally cost $3.75 for 5 pounds. If the price is reduced by 20%, how much will be saved on every 5 pounds?

______

22. How much will be saved on one pound of peaches?

______

SHARPEN YOUR SKILLS

Estimating a Percent of a Number

Estimate each answer.

1. 11% of 50 ________
2. 74% of 201 ________
3. 24% of 158 ________
4. 42% of 20 ________

5. 51% of 359 ________
6. 98% of 671 ________
7. 9% of 89 ________
8. 18% of 119 ________

9. 48% of 202 ________
10. 77% of 40 ________
11. 39% of 80 ________
12. 12% of 399 ________

Mental Math Exchange the percent and the number to do these mentally.

13. 44% of 50 ________
14. 16% of 75 ________
15. 28% of 25 ________
16. 48% of 75 ________

17. 86% of 50 ________
18. 35% of 20 ________
19. 64% of 25 ________
20. 5% of 80 ________

Use the table for Problems 21–23.
Estimate the number of tickets sold.

TYPES OF MOVIE TICKETS SOLD AT FILMFEST THEATER	
Full-price adults	48%
Children	23%
Senior citizens	29%
TOTAL TICKET SALES	
February	31,783
March	47,095

21. Full-price adult tickets sold in March ________

22. Children's tickets sold in February ________

23. Senior-citizen tickets sold in March ________

NAME ________________

SHARPEN YOUR SKILLS

Too Much or Too Little Information

Solve each problem about Luigi's Retaurant if possible. If there is too much information, write which facts are extra. If there is too little information, write what facts you would need to solve the problem.

1. Louis bought dinner at Luigi's Restaurant. The ravioli cost $6.00. Louis had lasagna for $7.00. How much did he pay in tax at 8%?

____

____

____

2. Two people sitting near Louis had the ravioli dinner. A third had ziti. How much was the total bill?

____

____

____

3. Antonio decided to order lasagna. How much did he pay including tax and a 15% tip on the price of the food?

____

____

____

4. What would the tax be on two ravioli dinners and one lasagna dinner, if the tax is 5%? The cost of dessert and salad is extra.

____

____

____

5. The Tompkins family ordered dinner for five. They had two spaghetti dinners at $5.00 each, two lasagna dinners, and one ravioli. How much did they pay for their dinner including a 15% tip and 6% tax?

____

____

6. After dinner, the three children wanted to buy t-shirts. The tax on the t-shirts was 6%. What was the total tax on the three shirts?

____

____

____